I0048781

Materials for Solar Cell Technologies II

Edited by

Inamuddin[1], Tauseef Ahmad Rangreez[2], Mohd Imran Ahamed[3] and Hamida-Tun-Nisa Chisti[2]

[1]Department of Applied Chemistry, Zakir Husain College of Engineering and Technology, Faculty of Engineering and Technology, Aligarh Muslim University, Aligarh-202 002, India

[2]Department of Chemistry, National Institute of Technology, Srinagar, Jammu and Kashmir 190006, India

[3]Department of Chemistry, Faculty of Science, Aligarh Muslim University, Aligarh 202 002, India

Copyright © 2021 by the authors

Published by **Materials Research Forum LLC**
Millersville, PA 17551, USA

All rights reserved. No part of the contents of this book may be reproduced or transmitted in any form or by any means without the written permission of the publisher.

Published as part of the book series
Materials Research Foundations
Volume 103 (2021)
ISSN 2471-8890 (Print)
ISSN 2471-8904 (Online)

Print ISBN 978-1-64490-140-3
eBook ISBN 978-1-64490-141-0

This book contains information obtained from authentic and highly regarded sources. Reasonable efforts have been made to publish reliable data and information, but the author and publisher cannot assume responsibility for the validity of all materials or the consequences of their use. The authors and publishers have attempted to trace the copyright holders of all material reproduced in this publication and apologize to copyright holders if permission to publish in this form has not been obtained. If any copyright material has not been acknowledged please write and let us know so we may rectify this in any future reprints.

Distributed worldwide by

Materials Research Forum LLC
105 Springdale Lane
Millersville, PA 17551
USA
https://www.mrforum.com

Manufactured in the United States of America
10 9 8 7 6 5 4 3 2 1

Table of Contents

Preface

Solar cells are idea choice energy-harvesting photovoltaic technology that can potentially meet the increasing global energy demands to unravel the fossil fuels energy crisis and environmental problems. Solar cells composed of varied semiconductor materials are arising all over the world to converting solar radiation into electricity by utilization of sunlight with zero greenhouse gas emissions. Expansion of materials discovery and production processes has acted a crucial part in solar cell technologies that have attracted global attention and extensive research in photovoltaics and energy conversion fields. However, there are many challenges before photovoltaics could provide clean, abundant, and cheap energy. Thus, the challenge of developing efficient and stable materials for emerging solar cell technologies has stimulated investigation of materials including conductive polymers, semiconductors, transition-metal compounds, alloys, and carbon materials, and among others. This book gives a comprehensive and unified summary of current researching progress and development trend of materials for solar cell technologies. Key topics include fabrication methods for solar energy materials and their utilization in various sorts of solar cell design and their merits and demerits, theoretical insights, and versatile applications in the present market. This book chronicles vital strides in solar cell device development research and benefit material scientists, professionals, faculty, postgraduate and graduate students operating in chemistry, physics, semiconductor technology, photochemistry, material science, light science, and nanotechnology.

Key features:

- The outlook for solar energy materials is described exploring the challenges faced by the energy industry
- Contains contributions from well-established experts in solar cell technology
- Encompasses practical solutions to present challenges in the solar cell industry
- Design to inform researchers and manufacturers of the recent research and developments in this rapidly growing field

Summary

Chapter 1 contains an introduction, past and present scenario of solar cell materials. The first section consists of the working principle whereas the second section deliberates the past scenario of solar cell materials in detail. Furthermore, in the next sections properties, different types, and present goals of these materials are discussed.

Chapter 2 discusses the concept of solar cells and its working principle. Solar cell offers several benefits. This chapter discussed in detail the applications of the solar cell in space research, telecommunication, grid connections, standalone devices, solar PV dryer, solar PV duster, off-grid power, solar aviation, and medical application.

Chapter 3 discusses the versatile applications of PEDOT:PSS as electrode, hole transport layer, and as a buffer layer in organic solar cells. In dye-sensitized solar cells, PEDOT:PSS can improve the catalytic activity of the counter electrode, whereas in silicon-based hybrid solar cells it can enhance the performances of the cell.

Chapter 4 provides an account of various types of transparent conducting electrode materials and their technologies applied in optoelectronic devices including the thin-film solar cells. In addition to the fundamental physics behind the transparent conductors, various nanostructured materials, including metallic nanowires, nano-carbon forms, and their integration as hybrid forms have been reviewed. A feasible road-map towards sustainable development of transparent conductors has been proposed.

Chapter 5 concisely deliberates various simulation models employed for modeling of photovoltaic materials along with its historic background. The main motivation of the chapter is the emphasis on the prominence of simulation models utilized in photovoltaic materials and the basic principles of materials modeling that are conferred in detail.

Chapter 6 addresses multiple applications and benefits of solar energy. Electricity can be generated either directly from photovoltaic cells or indirectly by raising steam in concentrating solar thermal systems. Solar energy is also exploited for hot water, space heating, cooling, and drying. Technology and policy breakthroughs should be forthcoming.

Chapter 7 discusses the various types of hybrid materials for solar cells consisting of carbon nanotubes, organic molecules, metal nanoparticles, polymer-nanoparticles, and their hybrid composites. The main focus is on the principle, efficiency, and working role of hybrid material in the solar cell.

Materials for Solar Cell Technologies II　　　　　　　　　　Materials Research Forum LLC
Materials Research Foundations **104** (2021) 1-23　　　　　https://doi.org/10.21741/9781644901410-1

Chapter 1

Introduction, Past and Present Scenario of Solar Cell Materials

M. Rizwan[1*], Waheed S. Khan[2], A. Khadija[3]

[1] School of Physical Sciences, University of the Punjab, Lahore, Pakistan

[2] Nanobiotechnology Group, National Institute for Biotechnology and Genetic Engineering (NIBGE), Jhang Road, Faisalabad-38000, Pakistan

[3] Department of Physics, University of Gujrat, Hafiz Hayat Campus, Gujrat City, Pakistan

*Corresponding Author: rizwan.sps@pu.edu.pk

Abstract

Solar cells convert sunlight into electricity directly. It is a reliable, non-toxic and pollution free source of electricity. Since 19th century researchers have been trying to investigate different materials for solar cell devices. Commercially, Si based solar are predominate in this field, however, with passage of time different materials have been reported. Solar cell techniques are based on three different generations. 1st generation is based on Si and 2nd generation includes thin-films of CuInGaSe, GaAs, CdTe and GaInP etc. whereas 3rd generation is based on organic, hybrid perovskites, quantum dot (QD)-sensitizers & dye-sensitizers solar cells. Among all these, the 3rd generation solar cells are the most efficient and more cost effective than 1nd and 2nd generation solar cells. The 2nd generation is less costly but also less efficient compared to 1st generation. 3rd generation faces degradation of the photovoltaic materials which is a major problem. In this chapter different reported materials since 19th century for solar cells are mentioned. The past and present scenarios of solar cells are discussed comprehensively. It is observed that Si-based and multijunction solar cells dominate the market. Although, theoretically it is reported that hybrid perovskites and quantum dot materials for solar cell are the most efficient materials for photovoltaic PV devices. In spite of the high efficiency the stability of organic, hybrid perovskites, QD-sensitizers &dye-sensitizer materials is a big challenge.

Keywords

Solar Cell, QD-Sensitisers, Dye-Sensitizers, Organic Materials, Hybrid Perovskites Materials

Contents

1.   Introduction

Solar cells generate electricity when these interact with sunlight by absorbing photons of particular energy wavelength. These are semiconductor materials and produce electricity in the presence of sunlight. Solar cells cannot be recharged like batteries. They produce electricity when sunlight interacts with them. When there is no light, the electricity stops. The generated electric power is measured in watts of kilowatts. A solar cell device is fabricated by using semiconductor material. These materials have completely or partially filled valance band (V.B) and empty or partially filled conduction band (C.B). When the sunlight with energy $>E_g$ falls on solar cell the bonds are brokken and loosely bonded electrons transfer from V.B to C.B. These electrons in the C.B are directed towards the external contact. These free electrons generate electricity in the external circuit. They transfer their energy to the external device and return towards V.B with the same initial energy. Figure 1 demonstrates the working principle of a solar cell. When the photons with energy >E_g (bandgap energy) interacts with solar cell it produces electrical power while photons with energy <E_g are absorbs in material and produce heat in the presence of direct sunlight at ambient temperature. Therefore, solar cells can operate at lower temperature and produce electricity without moving parts such as heat engine or generator to produce electricity.

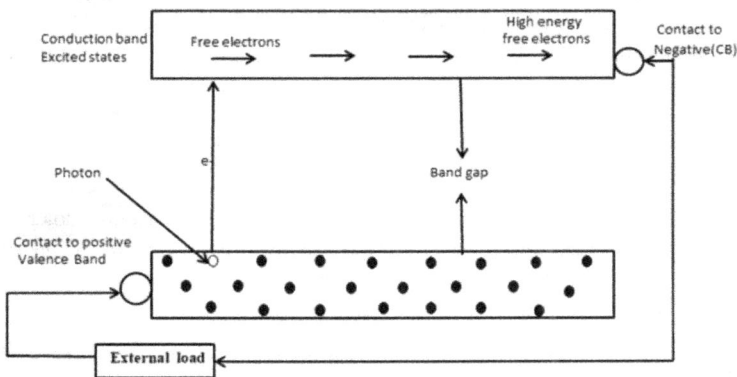

Figure 1 Schematic daigram of solar cell operation.

2.   Past scenario of solar cell materials

The history of photovoltaic (PV) materials dates back to the 19th century. In 1883, commercially for the first time a solar cell was introduced by Fritts [1]. He foresaw today's technological manufacturing and applications of solar cell devices over centuries ago. The year 1954 was the modern era of the solar cell industry. In 1954, in the USA at Bell Lab, the researchers discovered that pn-junctions based diodes produced voltage in the presence of room lights and stopped when lights were off. This provided a new field of research. After this, they had made Si based pn-junction solar cells within a year with efficiency of 6 % [2]. In the same year, a group of researchers in the USA reported heterojunction thin film based (Cu_2S/CdS) solar cell with the same efficiency 6 % [3] After one year, RCA Lab in USA reported GaAs based solar cells having efficiency of 6

Materials for Solar Cell Technologies II   Materials Research Forum LLC
Materials Research Foundations **104** (2021) 1-23   https://doi.org/10.21741/9781644901410-1

% [4]. Since 1960, several researchers published theoretical key papers on the pn-junction solar cells based on the relationship between thermodynamics, band gap, efficiency, incident light spectrum, and temperature [5]. Furthermore, CdTe thin film based solar cells were also reported with efficiency of 6 % [6]. Since 1960 the US satellites systems utilized Si solar cell technology in space for powering satellites [6]. Li-doped Si improved the radiation tolerant efficiency of existing Si based solar cell devices [7]. In 1970, a team of researchers in USSR at Ioffe institute made heteroface solar cells of GaAlAs/GaAs that pointed the new way of this device structures [8]. GaAs based solar cells had high resistance to ionize the radiations in outer space as well as had high efficiency. Therefore, GaAs based solar developed great interest towards solar cells. Its efficiency improved in both technical and non-technical levels in 1973. The 'violet cells' enhanced the efficiency to about 30% relative to Si cells [9]. GaAs heteroface structures were made in USA at IBM with 13% efficiency [9]. Moreover, the first world "Oil Embargo" agency was established by Persian oil producers to encourage the researchers for renewable energy devices especially photovoltaic devices in 1973. This revolutionized the industrial world [9].

In 1980, the industry started to work on the cost and manufacturing of solar materials. Manufacturing facilities for Si based pn-junction devices were present in the USA, Europe and Japan [10]. New technological schemes were built to scale up the solar cell technology based on thin films of a-Si and CuInSe$_2$ resultant solar cells achieved efficiency of >10%. These solar cells were made for small area (1 cm^2) by using controlled laboratory equipment [11]. It became more complicated for larger scale size due to insufficient facilities and equipment. Most the semiconductor companies in US gave their work. In 1990, the world largest company Arco solar worked in production of a-Si & n-Si thin films based PV devices and pre-commercialized thin films of CuInSe$_2$. In 1994 GaInP/GaAs multijunctons connector solar cells were made with efficiency of >30% [12]. Further, in 1998 thin films of Cu(InGa)Se$_2$ and triple-junctions solar cell (GaInP/GaAs/Ge) made by the US achieved high efficiency of > 19% [13]. Productions of CdTe and a-Si based thin films exceeded >100 MW/per year till 2002. In 2004, the efficiency of QDSSCs was reported of about 37% [14]. In 2007, the efficiency of thin-films solar cells based on Si (minimodule) was reported of >10% [15]. In 2012, the efficiency of GaAs materials was reported being 25% [16]. In 2013, efficiency of CZTSSe thin film based (thin film) solar cells was 12% [17]. After this in 2014 and 2016 the efficiency of CdTe materials was reported at 21% and 22% [18,19]. The efficiency of multicrytalline Si materials reached 22% in 2016 [20]. Moreover, till 2018 the efficiency of Si (crystalline) reached about >25% [21-24]. Recently, reported efficiency of Perovskites solar cell was 27.3% [25]. Furthermore, in 2019 the efficiency of organic solar cells reached about 16%. In spite of all these, low efficiency and stability of dye-sensitizer solar cells (DSSCs), quantum dot sensitizer solar cells (QDSSCs) and perovskites based solar materials is major a problem. Different reported crystalline, organic, dye and perovskite materials for solar cells are mentioned in table 1 [21,22,26-34].

Table 1 Different reported materials for solar cells.

S. No.	Classification	Efficiency %	Reference
	Crystalline		
1	Si	26.7 ± 0.5	[21]
		25.0 ± 0.5	[22]
		24.5 ± 0.5	[21]
		27.6 ± 1.2	[24]
		25.8 ± 0.5	[23]
2	InP	24.2 ± 0.5	[26]
	Multicrystalline		
3	Si	22.3 ± 0.4	[35]
4	GaAs	18.4 ± 0.5	[36]
	Thin films		
5	Si	10.5 ± 0.3	[15]
6		GaAs	[16]
7		21.0 ± 0.4	[18]
8	CdTe	18.6 ± 0.5	[37]
9		22.1 ± 0.5	[19]
10	CZTSSe	12.6 ± 0.3	[17]
11	CZTS	11.0 ± 0.2	[38]
12	GaAs	28.3 ± 0.9	[39]
13		23.3 ± 1.2	[40]
14	CIGS	10.0 ± 0.2	[41]
15		19.2 ± 0.5	[42]
16	GaInP	22.0 ± 0.3	[43]
	Multijunctions		
17	InGaP/GaAs/InGaAs	37.9 ± 1.2	[44]
18	GaInP/GaAs/Si	35.9 ± 0.5	[45]
19	a - Si/nc - Si/nc - Si	14.0 ± 0.4	[46]
20	GaInP/GaAs/Si	33.3 ± 1.2	[47]
21	GaInP/GaAs/Si	22.3 ± 0.8	[48]
22	GaAsP/Si	20.1 ± 1.3	[49]
23	GaAs/Si	32.8 ± 0.5	[45]
24	InGaP/GaAs/InGaAs	31.2 ± 1.2	[50]
25	AlGaInP/AlGaAs/GaAs/GaInAs	47.1 ± 2.6	[51]
26	a - Si/nc - Si	12.7 ± 0.4	[52]
27	GaInP/GaAs/GaInAs/GaInAs	45.7 ± 2.3	[53]

30	InGaP/GaAs/InGaAs	44.4 ± 2.6	[54]
31	GaInAsP/GaInAs	35.5 ± 1.2	[55]
	Perovskites		
32	Perovskite	24.2 ± 0.8	[30]
		20.9 ± 0.7	[31]
		17.25 ± 0.6	[29]
33	Organic	8.7 ± 0.3	[27]
		16.4 ± 0.2	[28]
34	Dye (cell)	11.9 ± 0.4	[33]
		10.7 ± 0.4	[32]
		8.8 ± 0.3	[34]

3. Properties of solar cell materials

Semiconductor materials are the potential materials for solar cells applications. These semiconductor materials have particular properties, which affect the efficiency of photovoltaic devices. Some of important properties are described as follows [56-58]:

- Material should have higher traping states or defects
- Direct-bandgap b/w 1.1 eVto 1.7 eV
- Non-toxic & easily reproducible technique
- Suitable to make larger area solar cell
- Stability for a long time
- Soluble in suitable solvent
- Higher conversion efficiency
- Smooth morphology

4. Categories of solar cell

There are three different generations regarding materials for solar cell devices. These are mentioned as follow:

4.1 First-generation solar cells

They are also named as Si-based solar cell generation. The highest efficiency achieved by single crystal solar cells is about 25%. But it is expensive to produce c-Si from sand and to deposit its single crystalline structure on another substrate. Expensive manufacturing of c-Si and deposition techniques motivated to develop new less costly materials and techniques for solar cell technology. Despite all of these c-Si still dominate over 90% of the market [56].

4.2   Second-generation solar cells

2nd generation is cheaper in manufacturing, fabrication techniques and materials as compared to 1st generation. These materials include nc-Si(nano-crystalline silicon), CuInGaSe, CdTe and amorphous Si(a-Si). These materials are cheaper than Si because they have higher absorption coefficient as compared to Si, which reduces the amount as well as cost of the material to be used. Only a single layer of these materials is enough to absorb a large amount of light and produce high power electricity. The efficiency of CuInGaSe, nc-Si, a-Si and CdTe is reported to be 20.1%, 10.1%, 9.5% and 16.7% respectively. This shows that although 2nd generation solar cells are cheaper in manufacturing than 1st generation solar cells they are less efficient in performance [56].

4.3   Third-generation solar cells

The target is to achieve higher efficiency as well as less costly manufacturing techniques as compared to the first two generations. The novelty in this generation was the use of semiconductor nano-particles. It expected that efficiency of photovoltaic (PV) devices can be achieved of about 31 to 41% by using semiconductor nano-particles due novel concepts of nano-particles like impact ionization [59,60]. Organic, perovskites hybrid, quantum dot & DSSCs solar cells are included in category of 3rd generation [61,62].

5.   Different materials for solar cells

Solar cell technology is based on semiconductor materials. Different materials for solar cells are described as follows:

5.1   Crystalline material for solar cells

Si in crystalline or multi-crystalline form is used frequently for solar cells applications. The 1st silicon based crystalline solar cell was developed by Chapin et al. [1] in 1954 with efficiency of >10% at Bell laboratories. After this it increased and reached >25% until 2017 [21-24]. The efficiency of multi-crystalline PV devices has reached about >22% in 2017 [15,36]. Conventional single crystalline materials were cost effective. In modern technology crystalline thin films, tri-crystalline solar cells were developed, which saved about 40% of material [63]. Although efficiency of polycrystalline or multi-crystalline Si based solar is lower than single crystal solar cells they are less costly than single crystal Si based solar cell. Single crystal Si based solar are expensive in fabrication. Furthermore, in place of Si different materials such as GaInP, CdTe, AlSb,GaN ,GaP, (CIS),GaSbInNCZTSSe, AlN, AlP, AlAs, GaAs,InAs, and InSb and, InP thin films enhanced the efficiency of solar cells [58,64].

5.2   Material for organic-solar cells

The semiconducting organic materials in the field of solar opened a new door of research. Organic polymer materials have many extinguished properties over silicon. They have high absorption coefficient about $\approx 10^5$ [65]. Only a thin layer of polymers about $\sim 100\ nm$ can absorb a large amount of light for energy conversion over silicon-based solar cells. They have tunable band-gap in visible range and yield large amount of charge carriers during solar cell operation [65]. Typically, organic polymers have lower band gap, larger active layers and high binding energy, which affect the performance of the solar cell. Organic solar cells consist of two different types, donor materials and acceptor materials.

First organic solar was developed by Ching Tang [66] in 1985 with efficiency 1%. He prepared it by the combination of perylene pigment and copper phthalocyanine (CuPc). It was a fully organic solar cell. After this in the mid 1990s another organic cell with efficiency >1% was reported by Messiner [67]. He prepared it by using C-60 doping with zinc phthalocyanine (ZnPc). After this several C-60 based hetero-junction solar were investigated [68-70]. Heerger et al. [65] developed polymer-fullerene based organic solar cell having efficiency of 6.5%. The donor reported polymers are poly(2-methoxy-5-(30,70-dimethyloctyloxy))-1,4-phenylenevinylene) (MDMO-PPV) and poly-3-hexyl-thiophene (P3HT), where acceptors are PCBM polymers and fullerene molecules. Recently non- fullerene based acceptor are also reported such as "perylene-diimide"(PDI), "naphthalene-diimide"(NDI), "diketopyr-rolopyrrole"(DPP) and "phthalimide (PhI)". These acceptors can also be used as potential candidates to achieve efficiency of solar cell up to ~ 13 to 14% [72-73]. But the average efficiency was between 8-10% till 2013 [74]. The limitation of organic materials for solar cell is their weak stability. They are unstable due to oxidation and degradation. To date efficiency of organic solar cells is about 10-15% [62].

5.3   Materials for DSSCs (dye-sensitized solar cells)

These materials move towards a new type of solar cells. The structure of DSSCs consists of four parts: electrolyte, photoanode, counter electrode and sensitizers as shown in figure 2. To improve the absorption rate of solar cells different dyes are used in DSSCs. Dyes are effective sensitizer to harvest photon up to the visible and IR region (400-800 nm) [56]. The first DDSCs was developed by Michel Gratzel et al. [75] in early 1990. The structure of TiO_2 nanocrystals with effective layer of molecule of dye had efficiency of about 12% [75-76]. TiO_3 in different morphologies such as rod, tubular, hierarchical and spherical shape affect the energy conversion efficiency of solar cell [56]. Nanocrystals of TiO_2 used as photoanode were coated on glass material such as FTO [56]. One of the reported materials for photoelectrodes is ZnO [77-79]. There are several other semiconductor oxide materials which can be used as photo-electrodes such as In_2O_3 [80-81], SnO_2 [79,80,82], $SrTiO_3$ [79,83,84], NiO [75, 79] and Nb_2O_5 [79,81,86,87]. Among them TiO_2 is the best candidate. It is more efficient than other materials. Performance of other oxides materials is yet to exceed the performance of TiO_2 nanocrystals [88-89]. Presently, the researchers are working on combine photoelectrodes such as SnO_2/ZnO to achieve high efficiency comparable to TiO_2. It was found that the efficiency of combined SnO_2/ZnO is greater than SnO_2 and ZnO on its own [90-91]. But to date TiO_2 based photoelectrodes are the most highly efficient. Zaban et al. [88] reported that efficiency of TiO_2 was improved by coating TiO_2 with Nb_2O_5. Different reported materials for dyes are Ru-based dyes such as "4-0-dicarboxylate)ruthenium(II) (N3 or N-719 dye)" and "cis-di(thiocyanato)bis(2,20-bipyridyl)-4", other compounds are "osmium (II) complex", "porphyrin dyes", "metal-free dyes", "porphyrin dyes" and "organic dyes (eosin yellow, 9-phenylxanthene, Perylene dye, Coumarin dye, Cyanine dye and Merocyanine dye")" [92-99]. Natural dyes can be extracted from leaves, fruits and colorful plants. The natural dyes have lower efficiency than artificial dyes. To the date Ru-based dyes are highly efficient. Different reported electrolyte material as Redox, iodine/tri-iodide (I_/I3_), SeCN_/(SeCN)$_3$ and Br_/Br3_, SCN_/ (SCN)$_3$ [100]. Among them Redox electrolyte showed the best performance [101]. The most commonly reported materials for counter electrodes are platinum and noble metals [56,102]. Other reported compounds are metal sulphide, transition metal nitrides, carbon-based materials, conducting polymers and carbides [102-104]. The efficiency of DSSCs is still reported $10 - 14\%$ [105].

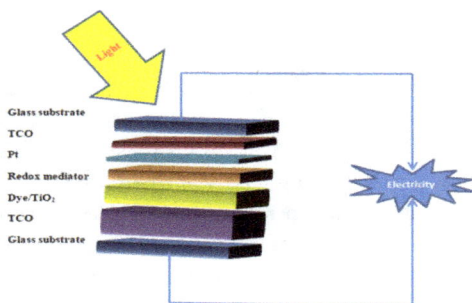

Figure 2 Schematic diagram of DSSC.

The limitations of DSSCs are their stability, contamination of photoelectrodes, degradation and desorption of dyes and leakage of electrolytes. Today researchers are working on to improve the efficiency of DSSCs.

5.4   Materials for QDSSCs

These materials have revolutionized the 3rd generation of solar cells. QDSSCs are based on quantum dots. The working principle of QDSSCs is similar to DSSCs. In QDSSCs dye is replaced by quantum dot. Quantum dot has tunable band gap properties, absorbs light up to IR region and produce multiple electrons to enhance efficiency of solar [14,106-109] cells. In recent years it is reported that quantum dots (QDs) of InAs with 5, 10 and 12 nm size have 1.071, 0.553 and 0.045 eV band gap respectively. The quantum dots at nano-scale harvest maximum amount of photons. Theoretically it was reported that for a single-junction solar cell device a single quantum dot could improve the energy conversion efficiency up to 31-44% [110]. In QDSSCs QDs are used as sensitizer with different photo-anodes. The different reported inorganic materials as QDs sensitizers are CdSe, CdTe, SnSe, CdS, PbSe, SnS and PbS. In recent years, the reported efficiency of QDSSCs is 8% [111]. Efficiency of QDSSCs can be affected by Pt-electrode. QDs are very sensitive to Pt. Researchers are working on Pt free electrode QDSSCs [56]. CdS-QDs-sensitized nanocrystalline under 1 sun illumination with TiO$_2$ had efficiency of 2.8% [112]. The assembly of CdSe-QDs along with CdS-QDs improved the absorption range from 550 (of CdS bulk) to 720 nm wavelength (of CdSe bulk) [112]. It improved efficiency up to 4.2% under one illumination. Assembly of quantum dots with different polymers can improve efficiency of solar cells [112].

5.5   Materials for hybrid perovskites solar cells

Hybrid perovskites follow the ABX3 structure in which A is an organic cations ("CH$_3$NH$_3$+, C$_2$H$_5$NH$_3$"), B is any metal cation ("Pb^{2+}, Sn^{2+}, Bi^{2+}", etc.) and X is any Halide anion"(Cl$^-$, Br$^-$, I$^-$)". The class of perovskites is used as potential candidate in solar cell application. In 2009, the reported efficiency of first perovskite based solar was 3.81% [113]. It was reported that halide perovskites sensitizer showed higher absorption coefficient in visible range than other reported organic dyes. The perovskite "CH$_3$NH$_3$PbI$_3$" showed excellent ability to harvest the photons. Since 2009 to 2014, the recorded efficiency of "CH$_3$NH$_3$PbI$_3$"-based solar cell was increased from 3.8 to 16.7% [113-117]. The properties of perovskites could be tuned to improve efficiency of PV devices. Different materials such as "PEDOT:PSS", "spiro-OMeTAD", "poly-triarylamine (PTAA)", "PCPDTBT", "P$_3$HT", "PCDTBT"and"CsSnI$_3$" were investigated with CH$_3$NH$_3$PbI$_3$ in PV device applications [56]. Their efficiency improved over DSSCs. It would be expected that it could be increased up to >25% in near future.

Recently reported efficiency was 27.3% [25]. In $CH_3NH_3PbI_3$ perovskite Pb is a toxic material. Researchers focused on lead free perovskites materials to enhance the device efficiency. In place of Pb reported material were stibium (Sb), bismuth (Bi) and tin (Sn). Recently, the efficiency of $FASnI_3$ perovskites has reached up to 9%. The stability of Bi based perovskites materials is higher than lead based materials. In 2018, the reported efficiency of Bi-based perovskites has reached about 3.17% [56]. Next target of hybride PSCs is to enhance the efficiency of lead free perovskites solar cells.

6.   Present goal for solar cell materials

Currently most studied materials for solar cell applications are silicon and III-V group compounds. The efficiency of thin films or multijunctions solar cells of GaInP, GaAs, and Ge layers is reported up to 30%. Si is still a 99% commercial dominated material for PV devices. The concept of 3rd generation is under investigation. It does not overtake the market. The Commercially reported efficiency of 1st generation and 2nd generation solar cells is higher than 3^{rd}generation solar cells. Dye-sensitizer (DS) solar cells, hybrid perovskites solar cells and quantum dot (QD) solar cells are less costly and have high efficiency but these are not yet commercially fabricated for solar cell applications. The stability factor of organic and inorganic sensitizers affects their efficiency. The hardly achieved efficiencies of DSSCs, QDSSCs, organic solar cell and hybrid-perovskites solar cell are 10-14, 10, 8-10 and 15% respectively [56]. Commercially Si based and multijuction solar cells achieved the highest efficiency up to 30% [14,37]. Commercially high efficient materials are more preferable than 3rd generation materials. Still the present research is to enhance the efficiency of multijunction solar devices such as"InGaP/GaAs/InGaAs,InGaP/GaAs/InGaAs,GaInP/GaAs/Si(waferbonded), GaInP/GaAs/Si (monolithic), AlGaInP/AlGaAs/GaAs/GaInAs, GaInP/GaAs/GaInAsP/GaInAs, GaInAsP/GaInAs, aSi/nc-Si/nc-Si and InGaP/GaAs/InGaAs" etc. [64]. There is need to investigate more stable and less costly materials for 3rd generation solar cells to gain higher efficiency.

Conclusions

All the discussion concluded that solar cell opened a new field of research. With increasing demand of solar efficiency and less costly fabrication researchers are working on different materials for solar cell devices and its applications. Commercially Si-based solar are high efficient with about >30%. Si-based technology attracted the attention of researchers. It showed high efficiency and stability. Different multijuncton solar cells materials such as "InGaP/GaAs/InGaAs,InGaP/GaAs/InGaAs, GaInP/GaAs/Si),GaInP/GaAs/Si(monolithic), AlGaInP/AlGaAs/GaAs/GaInAs,

GaInP/GaAs/GaInAsP/GaInAs, GaInAsP/GaInAs, amorphous-Si/nanocrystaline-Si/nanocrystaline-Si" had efficiency comparable to silicon. To the date average reported efficiency for organic, perovskites, dye-sensitizer and quantum dot sensitizer materials is 10, 15, 8-10 and 10% respectively. Due to degradation and environmental factors the stability of organic, perovskites, dye-sensitizer and quantum dot sensitizer materials is a big challenge. Theoretically, reported efficiency for quantum dot sensitizers is about >60% and perovskites >30%. Researchers are trying to discover high stable materials in the near future for the sake of potential applications of solar cell devices.

Acknowledgements

Authors are immensely grateful to Department of Physics, University of Gujrat, Gujrat-50700, Pakistan and National Institute for Biotechnology & Genetic Engineering (NIBGE), Faisalabad-38000, Pakistan for providing healthy and conducive environment for such work.

References

[1] C.E. Fritts, On a new form of selenium cell, and some electrical discoveries made by its use, Am. J. Sci. 26 (1883) 465-472. https://doi.org/10.2475/ajs.s3-26.156.465

[2] D.M. Chapin, C. Fuller, G. Pearson, A new silicon p-n junction photocell for converting solar radiation into electrical power, J. Appl. Phys. 25 (1954) 676-677 https://doi.org/10.1063/1.1721711

[3] D. Reynolds, G. Leies, L. Antes, R. Marburger, Photovoltaic effect in cadmium sulfide, Ph.ys Rev. 96 (1954) 533-534. https://doi.org/10.1103/PhysRev.96.533

[4] D. Jenny, J. Loferski, P. Rappaport, Photovoltaic effect in GaAs p– n junctions and solar energy conversion, Phys. Rev. 101 (1956) 1208-1209. https://doi.org/10.1103/Phys Rev.101.1208

[5] M. Prince, Silicon solar energy converters, J. Appl. Phys. 26 (1955) 534-540. https://doi.org/10.1063/1.1722034

[6] D. Cusano, CdTe solar cells and photovoltaic heterojunctions in II–VI compounds, Solid State Electron. 6 (1963) 217-232 https://doi.org/10.1016/0038-1101(63)90078-9

[7] J. Wysocki, Lithium-doped radiation-resistant silicon solar cells, IEEE T Nucl. Sci. 13 (1966) 168-173. https://doi.org/10.1109/TNS.1996.4324358

[8] Z.I. Alferov, V. Andreev, M. Kagan, I. Protasov, V. Trofim, Solar-energy converters based on pn Al_xGa_{1-x} As-GaAs heterojunctions, Sov. Phys.-Semicond.(Engl. Transl.);(United States). 4 (1971) 16-25. https://doi.org/10.1117/12.934271.

[9] J. Lindmayer, J. Allison, The violet cell: an improved silicon solar cell. Sol. Cells 29 (1990) 151-166 https://doi.org/10.1016/0379-6787(90)90023-X.

[10] S. S. Hegedus, A. Luque, Status, trends, challenges and the bright future of solar electricity from photovoltaics. Handbook of photovoltaic science and engineering, John Wiley(2003)1-43. https://doi.org/10.1002/0470014008

[11] R.Sinton, Y. Kwark, J. Gan, R.M. Swanson, 27.5-percent silicon concentrator solar cells. IEEE Electron Device Lett.7 (1986) 567-569. https://doi.org/10.1109/EDL.1986.26476

[12] D.Friedman, S.R. Kurtz, K. Bertness, A. Kibbler, C. Kramer, J. Olson, D. King, B. Hansen, J. Snyder, Accelerated publication 30.2% efficient GaInP/GaAs monolithic two terminal tandem concentrator cell. Prog. Photovolt. 3 (1995) 47-50. https://doi.org/10.1002/pip.4670030105

[13] M.A. Contreras, B. Egaas, K. Ramanathan, J. Hiltner, A. Swartzlander, F. Hasoon, R. Noufi, Progress toward 20% efficiency in Cu (In, Ga) Se_2 polycrystalline thin film solar cells. Prog. Photovolt. 7 (1999) 311-316. https://doi.org/10.1002/pip.4670030105

[14] R. King, C. Fetzer, K. Edmondson, D. Law, P. Colter, H. Cotal, R. Sherif, H. Yoon, T. Isshiki, D. Krut. Metamorphic III-V materials, sublattice disorder, and multijunction solar cell approaches with over 37% efficiency, 19th European photovoltaic solar energy conference and exhibition. (2004) 7-11.

[15] M.J. Keevers, T.L. Young, U. Schubert, M.A. Green. 10% efficient CSG minimodules, 22nd European photovoltaic solar energy conference, (2007) 1783-1790.

[16] L.S. Mattos, S.R. Scully, M. Syfu, E. Olson, L. Yang, C. Ling, B.M. Kayes, G. He. New module efficiency record: 23.5% under 1-sun illumination using thin-film single-junction GaAs solar cells. in 2012 38th IEEE photovoltaic specialists conference. 38 (2012) 1-4. https://doi.org/10.1109/PVSC.2012.6318255

[17] W. Wang, M.T. Winkler, O. Gunawan, T. Gokmen, T.K. Todorov, Y. Zhu, D.B. Mitzi, Device characteristics of CZTSSe thin film solar cells with 12.6% efficiency. Adv. Energy Mater. 4 (2014) 1-5. https://doi.org/10.1002/aenm.201301465

[18] F. Solar, First solar builds the highest efficiency thin film PV cell on record, 2014.

[19] M. Tuteja, A.B. Mei, V. Palekis, A. Hall, S. MacLaren, C.S. Ferekides, A.A. Rockett, $CdCl_2$ treatment-induced enhanced conductivity in CdTe solar cells observed

using conductive atomic force microscopy, J. Phys. Chem. Lett. 7 (2016) 4962-4967. https://doi.org/10.1021/acs.jpclett.6b02399

[20] P. Verlinden, Will we have> 22% efficient multi crystalline silicon solar cells, PVSEC. 26 (2016) 24-28.

[21] K. Yoshikawa, H. Kawasaki, W. Yoshida, T. Irie, K. Konishi, K. Nakano, T. Uto, D. Adachi, M. Kanematsu, H. Uzu, Silicon heterojunction solar cell with interdigitated back contacts for a photoconversion efficiency over 26%, Nat. Energy. 2 (2017) 17-32. https://doi.org/10.1038/nenergy.2017.32

[22] J. Zhao, A. Wang, M.A. Green, F. Ferrazza, 19.8% efficient "honeycomb" textured multicrystalline and 24.4% monocrystalline silicon solar cells. Appl. Phys. Lett.73 (1998) 1991-1993. https://doi.org/10.1063/1.122345

[23] F. Haase, C. Hollemann, S. Schaefer, A. Merkle, M. Rienaecker, J. Krügener, R. Brendel, R. Peibst, Laser contact openings for local poly-Si-metal contacts enabling 26.1%-efficient POLO-IBC solar cells. Sol. Energy Mater. Sol. 186(2018) 184-193. https://doi.org/10.1016/j.solmat.2018.06.020.

[24] A. Richter, J. Benick, F. Feldmann, A. Fell, M. Hermle, and S.W. Glunz, n-Type Si solar cells with passivating electron contact: Identifying sources for efficiency limitations by wafer thickness and resistivity variation. Sol. Energy. Mater. Sol. 173 (2017) 96-105. https://doi.org/10.1016/j.solmat.2017.05.042

[25] P. Oxford, Oxford PV sets world record for perovskite solar cell. 2018

[26] M. Wanlass, Systems and methods for advanced ultra-high-performance InP solar cells: Google Patents. 2017.

[27] M. Hosoya, H. Oooka, H. Nakao, T. Gotanda, S. Mori, N. Shida, R. Hayase, Y. Nakano, M. Saito. Organic thin film photovoltaic modules. in Proceedings of the 93rd Annual Meeting of the Chemical Society of Japan. 2013.

[28] A. Laventure, C.R. Harding, E. Cieplechowicz, Z. Li, J. Wang, Y. Zou, G.C. Welch, Screening quinoxaline-type donor polymers for roll-to-roll processing compatible organic photovoltaics. ACS Appl. Polym. Mater. 1 (2019) 2168-2176. https://doi.org/10.1021/acsapm. 9b00433.

[29] K. Unnikrishnan, Environmental chamber to regulate film morphology for solar energy materials printing using additive manufacturing and investigating the role of additives in perovskites, University of Washington Libraries, 2018.

[30] E.H. Jung, N.J. Jeon, E.Y. Park, C.S. Moon, T.J. Shin, T.-Y. Yang, J.H. Noh, J. Seo, Efficient, stable and scalable perovskite solar cells using poly (3-hexylthiophene). Nat. 567 (2019) 511. https://doi.org/10.1038/s41586-019-1036-3

[31] W.S. Yang, J.H. Noh, N.J. Jeon, Y.C. Kim, S. Ryu, J. Seo, S.I. Seok, High-performance photovoltaic perovskite layers fabricated through intramolecular exchange. sci. 348 (2015) 1234-1237. https://doi.org/10.1126/science.aaa9272

[32] M. Kawai, High-durability dye improves efficiency of dye-sensitized solar cells. Nikkei Electronics, (2013).

[33] R. Komiya, A. Fukui, N. Murofushi, N. Koide, R. Yamanaka, H. Katayama. Improvement of the conversion efficiency of a monolithic type dye-sensitized solar cell module. in Technical Digest, 21st International photovoltaic science and engineering conference. 2011.

[34] S. Mori, H. Oh-oka, H. Nakao, T. Gotanda, Y. Nakano, H. Jung, A. Iida, R. Hayase, N. Shida, M. Saito, Organic photovoltaic module development with inverted device structure. Mater. Res. Soc. Symp. Proc. 1737 (2015) 27-15. https://doi.org/10.1557/opl.2015.540

[35] J. Benick, A. Richter, R. Müller, H. Hauser, F. Feldmann, P. Krenckel, S. Riepe, F. Schindler, M.C. Schubert, M. Hermle, High-efficiency n-type HP mc silicon solar cells. IEEE J. Photovolt. 7 (2017) 1171-1175. https://doi.org/10.1109/JPHOTOV.2017.2714139.

[36] R. Venkatasubramanian, B. O'Quinn, J. Hills, P. Sharps, M. Timmons, J. Hutchby, H. Field, R. Ahrenkiel, B. Keyes. 18.2% (AM1. 5) efficient GaAs solar cell on optical-grade polycrystalline Ge substrate in conference record of the twenty fifth IEEE photovoltaic specialists conference. 25 (1996) 31-36. https://doi.org/10.1109/PVSC.1996.563940

[37] M.A. Green, Commercial progress and challenges for photovoltaics. Nat Energy. 1 (2016) 1-4. https://doi.org/10.1038/nenergy.2015.15

[38] K. Sun, C. Yan, F. Liu, J. Huang, F. Zhou, J.A. Stride, M. Green, X. Hao, Over 9% efficient kesterite Cu_2ZnSnS_4 solar cell fabricated by using Zn1−xCdxS buffer layer. Adv. Energy Mater. 6 (2016) 160-46. https://doi.org/10.1002/aenm.201600046

[39] M.A. Green, Y. Hishikawa, W. Warta, E.D. Dunlop, D.H. Levi, J. Hohl Ebinger, A.W. HoBaillie, Solar cell efficiency tables (version 50). Prog Photovolt. 25 (2017) 668-676. https://doi.org/10.1002/pip.2909

[40] M. Alves, A. Pérez-Rodríguez, P.J. Dale, C.D. Domínguez, S. Sadewasser, Thin-film micro-concentrator solar cells. J. Phys. Energy. 02 (2019) 76-55. https://doi.org/10.1088/2515-7655/ab4289

[41] C.Yan, C., J. Huang, K. Sun, S. Johnston, Y. Zhang, H. Sun, A. Pu, M. He, F. Liu, K. Eder, $Cu_2 ZnSnS_4$ solar cells with over 10% power conversion efficiency enabled by heterojunction heat treatment. Nat. Energy. 3 (2018) 764–772. https://doi.org/10.1038/s41560-018-0206-0

[42] H. Sugimoto, High efficiency and large volume production of CIS-based modules. in 2014 IEEE 40th Photovoltaic specialist conference (PVSC).40 (2014) 2767-2770. https://doi.org/10.1109/PVSC.2014.6925503.

[43] M.P. Van der Laan, S.J. Andersen, P.]E. Réthoré, Brief communication: Wind speed independent actuator disk control for faster AEP calculations of wind farms using CFD. Wind energy science discussions, 2019.

[44] A.W. Bett, S.P. Philipps, S. Essig, S. Heckelmann, R. Kellenbenz, V. Klinger, M. Niemeyer, D. Lackner, F. Dimroth. Overview about technology perspectives for high efficiency solar cells for space and terrestrial applications. in 28th European photovoltaic solar energy conference and exhibition. 2013.

[45] S. Essig, C. Allebé, T. Remo, J.F. Geisz, M.A. Steiner, K. Horowitz, L. Barraud, J.S. Ward, M. Schnabel, A. Descoeudres, Raising the one-sun conversion efficiency of III–V/Si solar cells to 32.8% for two junctions and 35.9% for three junctions. Nat. Energy. 2 (2017) 1-8. https://doi.org/10.1038/nenergy.2017.144

[46] H. Sai, T. Matsui, K. Matsubara, Stabilized 14.0%-efficient triple-junction thin-film silicon solar cell. App. Phys. Lett. 109 (2016) 183-506. https://doi.org/10.1063/1.4966996

[47] R. Cariou, J. Benick, F. Feldmann, O. Höhn, H. Hauser, P. Beutel, N. Razek, M. Wimplinger, B. Bläsi, D. Lackner, III–V-on-silicon solar cells reaching 33% photoconversion efficiency in two-terminal configuration. Nat. Energy. 3 (2018) 326–333. https://doi.org/10.1038/s41560-018-0125-0

[48] M. Feifel, J. Ohlmann, J. Benick, M. Hermle, J. Belz, A. Beyer, K. Volz, T. Hannappel, A.W. Bett, D. Lackner, Direct growth of III–V/silicon triple-junction solar cells with 19.7% efficiency. IEEE J. Photovolt. 8 (2018) 1590-1595. https://doi.org/10.1109/JPHOTOV. 2018.2868015

[49] T.J. Grassman, D.J. Chmielewski, S.D. Carnevale, J.A. Carlin, S.A. Ringel. GaAsP/Si dual-junction solar cells grown by MBE and MOCVD. in 2015 IEEE 42nd

Photovoltaic specialist conference (PVSC). (2015) 1-5.
https://doi.org/10.1109/PVSC.2015.7356384.

[50] T. Takamoto, H. Washio, H. Juso. Application of InGaP/GaAs/InGaAs triple junction solar cells to space use and concentrator photovoltaic. in 2014 IEEE 40th photovoltaic specialist conference (PVSC). (2014) 0001-0005.
https://doi.org/10.1109/PVSC.2014.6924936.

[51] J.F. Geisz, M.A. Steiner, N. Jain, K.L. Schulte, R.M. France, W.E. McMahon, E.E. Perl, D.J. Friedman, Building a six-junction inverted metamorphic concentrator solar cell. IEEE J. Photovolt. 8 (2017) 626-632.
https://doi.org/10.1109/JPHOTOV.2017.2778567

[52] T. Matsui, K. Maejima, A. Bidiville, H. Sai, T. Koida, T. Suezaki, M. Matsumoto, K. Saito, I. Yoshida, M. Kondo, High-efficiency thin-film silicon solar cells realized by integrating stable a-Si: H absorbers into improved device design. Japanese J. Appl. Phys. 54 (2015)1-5. https://doi.org/10.7567/JJAP.54.08KB10

[53] Lin,Chu-Jian Investigation of Gallium Arsenide Thin Films deposited by RF Sputtering, NCU Institutional Repository.

[54] H. Lv, F. Sheng, J. Dai, W. Liu, C. Cheng, and J. Zhang, Temperature-dependent model of concentrator photovoltaic modules combining optical elements and III–V multi-junction solar cells. Sol. Energy. 112 (2015) 351-360.
https://doi.org/10.1016/j.solener.2014.12.005

[55] N. Jain, K.L. Schulte, J.F. Geisz, D.J. Friedman, R.M. France, E.E. Perl, A.G. Norman, H.L. Guthrey, M.A. Steiner, High-efficiency inverted metamorphic 1.7/1.1 eV GaInAsP/GaInAs dual-junction solar cells. App. Phys. Lett. 112 (2018) 0539051-0539055. https://doi.org/10.1063/1.5008517

[56] S. Ananthakumar, J.R. Kumar, S.M. Babu, Third-generation solar cells: concept, materials and performance-an overview, in emerging nanostructured materials for energy, Environ. Sci. (2019) 305-339. https://doi.org/10.1007/978-3-030-04474-9_7

[57] I. Dharmadasa, Advances in thin-film solar cells: Jenny Stanford Publishing. (2013). https://doi.org/10.1201/9780429020841

[58] G. Adolf, H. Christopher, Photovoltaic materials, past, present, future. Sol. Energy Mater. Sol. 62 (2000) 1-19. https://doi.org/10.1016/B978-185617390-2/50006-4

[59] K.H. Yoon, Beard trimmer: Google Patents,2014.

[60] L.R.Weiss, S.L. Bayliss, F. Kraffert, K.J. Thorley, J.E. Anthony, R. Bittl, R.H. Friend, A. Rao, N.C. Greenham, J. Behrends, Strongly exchange-coupled triplet pairs in an organic semiconductor. Nat Phys.13 (2017)1-5. https://doi.org/10.1038/nphys3908

[61]   F.W. Low C.W. Lai, Recent developments of graphene-TiO_2 composite nanomaterials as efficient photoelectrodes in dye-sensitized solar cells: A review. Renew. Sust. Energ. Rev. 82 (2018) 103-125. https://doi.org/10.1016/j.rser.2017.09.024

[62] Drevets, W.C. Q.S. Li, Method for the treatment of depression: Google Patents.2018.

[63] A.Goetzberger C. Hebling, Photovoltaic materials, past, present, future. Sol. Energy Mater. Sol. 62 (2000) 1-19. https://doi.org/10.1016/S0927-0248(99)00131-2

[64] M.A.V. Green, E.D. Dunlop, D.H. Levi, J. Hohl Ebinger, M. Yoshita, A.W. Ho Baillie, Solar cell efficiency tables (version 54). Prog Photovolt. 27(2019) 565-575. https://doi.org/10.1002/pip.3171

[65] S. Günes, H. Neugebauer, N.S. Sariciftci, Conjugated polymer-based organic solar cells. Chem Rev. 107 (2007) 1324-1338. https://doi.org/10.1021/cr050149z

[66] C.W. Tang, Two layer organic photovoltaic cell. App. Phys. Lett. 48 (1986) 183-185. https://doi.org/10.1063/1.96937

[67] M. Westphalen, U. Kreibig, J. Rostalski, H. Lüth, D. Meissner, Metal cluster enhanced organic solar cells. Sol. Energy. Mater. Sol. 61(2000) 97-105. https://doi.org/10.1016/S0927-0248(99)00100-2

[68] N. Sariciftci, D. Braun, C. Zhang, V. Srdanov, A. Heeger, G. Stucky, F. Wudl, Semiconducting polymer buckminsterfullerene heterojunctions: Diodes, photodiodes, and photovoltaic cells. App. Phys. Lett. 62 (1993) 585-587. https://doi.org/10.1063/1.108863

[69] B. Kraabel, D. McBranch, N. Sariciftci, D. Moses, A. Heeger, Ultrafast spectroscopic studies of photoinduced electron transfer from semiconducting polymers to C 60. Phys. Rev. B. 50 (1994) 18543. https://doi.org/10.1103/PhysRevB.50.18543

[70] N.Sariciftci, F. Wudl, A. Heeger, M. Maggini, G. Scorrano, M. Prato, J. Bourassa, P. Ford, Photoinduced electron transfer and long lived charge separation in a donor-bridge-acceptor supramolecular 'diad'consisting of ruthenium (II) tris (bipyridine) functionalized C60. Chem. Phys. lett. 247 (1995) 510-514. https://doi.org/10.1016/S0009-2614(95)01276-1.

[71] J.Y. Kim, K. Lee, N.E. Coates, D. Moses, T.Q. Nguyen, M. Dante, A.J. Heeger, Efficient tandem polymer solar cells fabricated by all-solution processing. Science 317 (2007) 222-225. https://doi.org/10.1126/science.1141711

[72] T. Chang, Real analysis 2 (2018) 1-5.

[73] E.J. Benjamin, S.S. Virani, C.W. Callaway, A.M. Chamberlain, A.R. Chang, S. Cheng, S.E. Chiuve, M. Cushman, F.N. Delling, R. Deo, Heart disease and stroke statistics-2018 update: a report from the Circ. Res. 137 (2018) 67-426 https://doi.org/10.1161/CIR.00000 00000000558

[74] M.C. Scharber N.S. Sariciftci, Efficiency of bulk-heterojunction organic solar cells. Prog Polym. Sci. 38 (2013)1929-1940. https://doi.org/10.1016/j.progpolymsci.2013.05.001

[75] M. Gratzel B. O'Regan, A low-cost, high-efficiency solar cell based on dye-sensitized colloidal TiO_2 films. Nat. 353 (1991) 737-740. https://doi.org/10.1021/nl400286w

[76] S. Ito, M.K. Nazeeruddin, P. Liska, P. Comte, R. Charvet, P. Péchy, M. Jirousek, A. Kay, S.M. Zakeeruddin, M. Grätzel, Photovoltaic characterization of dye sensitized solar cells: effect of device masking on conversion efficiency. Prog. photovolt. 14 (2006) 589-601 https://doi.org/10.1002/pip.683

[77] G. Redmond, D. Fitzmaurice, M. Graetzel, Visible light sensitization by cis-bis (thiocyanato) bis (2, 2'-bipyridyl-4, 4'-dicarboxylato) ruthenium (II) of a transparent nanocrystalline ZnO film prepared by sol-gel techniques. Chem Mater. 6 (1994) 686-691. https://doi.org/10.1021/cm00041a020.

[78] Y. Tachibana, K. Hara, K. Sayama, H. Arakawa, Quantitative analysis of light-harvesting efficiency and electron-transfer yield in ruthenium-dye-sensitized nanocrystalline TiO_2 solar cells. Chem. Mater. 14 (2002) 2527-2535. https://doi.org/10.1021/cm 011563s

[79] K. Eguchi, H. Koga, K. Sekizawa, K. Sasaki, Nb2O5-based composite electrodes for dye-sensitized solar cells. J. Ceram. Soc. Japan. 108 (2000) 1067-1071. https://doi.org/10.2109/jcersj.108.1264_1067

[80] H. Rensmo, K. Keis, H. Lindström, S. Södergren, A. Solbrand, A. Hagfeldt, S.-E. Lindquist, L. Wang, M. Muhammed, High light-to-energy conversion efficiencies for solar cells based on nanostructured ZnO electrodes. J. Phys. Chem. B. 101 (1997) 2598-2601. https://doi.org/10.1021/jp962918b

Materials for Solar Cell Technologies II   Materials Research Forum LLC
Materials Research Foundations **104** (2021) 1-23   https://doi.org/10.21741/9781644901410-1

[81] K. Hara, T. Horiguchi, T. Kinoshita, K. Sayama, H. Sugihara, H. Arakawa, Highly efficient photon-to-electron conversion with mercurochrome-sensitized nanoporous oxide semiconductor solar cells. Sol. Energy Mater. Sol. 64 (2000) 115-134 https://doi.org/1016/S0927-0248(00)00065-9

[82] N.N. Dinh, M.-C. Bernard, A. Hugot-Le Goff, T. Stergiopoulos, P. Falaras, Photoelectrochemical solar cells based on SnO_2 nanocrystalline films. Cr. Chim. 9 (2006) 676-683. https://doi.org/10.1016/j.crci.2005.02.042

[83] S.Gholamrezaei, M.S. Niasari, M. Dadkhah, B. Sarkhosh, New modified sol–gel method for preparation $SrTiO_3$ nanostructures and their application in dye-sensitized solar cells. J. Mater. Sci. Mater. 27 (2016) pp. 118-125. https://doi.org/10.1007/s10854-015-3726-4

[84] S.Burnside, J.-E. Moser, K. Brooks, M. Grätzel, D. Cahen, Nanocrystalline mesoporous strontium titanate as photoelectrode material for photosensitized solar devices: increasing photovoltage through flatband potential engineering. J. Phys. Chem. B.103 (1999) 9328-9332. https://doi.org/10.1021/jp9913867

[85] J. He, H. Lindström, A. Hagfeldt, S.-E. Lindquist, Dye-sensitized nanostructured p-type nickel oxide film as a photocathode for a solar cell. J. Phys Chem. B. 103 (1999) 8940-8943 https://doi.org/10.1021/jp991681r.

[86] K. Sayama, H. Sugihara, H. Arakawa, Photoelectrochemical properties of a porous Nb_2O_5 electrode sensitized by a ruthenium dye. Chem. Mater. 10 (1998) 3825-3832. https://doi.org/10.1021/cm980111l.

[87] T.N. Rao, L. Bahadur, Photoelectrochemical studies on dye sensitized particulate ZnO thin film photoelectrodes in nonaqueous media. J. Electrochem. Soc. 144 (1997) 179-185. https://doi.org/10.1149/1.1837382

[88] A. Zaban, S. Chen, S. Chappel, B. Gregg, Bilayer nanoporous electrodes for dye sensitized solar cells. Chem. Comm. 22 (2000) 2231-2232. https://doi.org/10.1039/B005921H

[89] A. Kay, M. Graetzel, Artificial photosynthesis. 1. Photosensitization of titania solar cells with chlorophyll derivatives and related natural porphyrins. J. Phys. Chem. 97 (1993) 6272-6277. https://doi.org/10.1021/j100125a029

[90] V.A.S. Perera, An efficient dye-sensitized photoelectrochemical solar cell made from oxides of tin and zinc. Chem. Comm. (1999) 15-16 https://doi.org/10.1039/A806801A

Materials Research Forum LLC
https://doi.org/10.21741/9781644901410-1

[91] K. Tennakone, G. Senadeera, V. Perera, I. Kottegoda, L. De Silva, Dye-sensitized photoelectrochemical cells based on porous SnO_2/ZnO composite and TiO_2 films with a polymer electrolyte. Chem Mater. 11 (1999) 2474-2477. https://doi.org/10.1021/cm990165a

[92] S. Ferrere, A. Zaban, B.A. Gregg, Dye sensitization of nanocrystalline tin oxide by perylene derivatives. J. Phys. Chem. B. 101 (1997) 4490-4493 https://doi.org/10.1021/jp970683d

[93] K. Sayama, M. Sugino, H. Sugihara, Y. Abe, H. Arakawa, Photosensitization of porous TiO_2 semiconductor electrode with xanthene dyes. Chem Lett. 27 (1998) 753-754. https://doi.org/10.1246/cl.1998.753

[94] A.C. Khazraji, S. Hotchandani, S. Das, P.V. Kamat, Controlling dye (Merocyanine-540) aggregation on nanostructured TiO_2 films. An organized assembly approach for enhancing the efficiency of photosensitization. J. Phys. Chem B. 103 (1999) 4693-4700. https://doi.org/10.1021/jp9903110.

[95] Z.S. Wang, F.Y. Li, C.H. Huang, Highly efficient sensitization of nanocrystalline TiO_2 films with styryl benzothiazolium propylsulfonate. Chem. Comm. 20 (2000) 2063-2064. https://doi.org/10.1039/b006427k.

[96] Z.S. Wang, F.Y. Li, C.H. Huang, L. Wang, M. Wei, L.P. Jin, N.Q. Li, Photoelectric conversion properties of nanocrystalline TiO_2 electrodes sensitized with hemicyanine derivatives. J. Phys. Chem. B. 104 (2000)9676-9682. https://doi.org/10.1021/jp001580p.

[97] K. Sayama, K. Hara, N. Mori, M. Satsuki, S. Suga, S. Tsukagoshi, Y. Abe, H. Sugihara, H. Arakawa, Photosensitization of a porous TiO_2 electrode with merocyanine dyes containing a carboxyl group and a long alkyl chain. Chem. Comm.13 (2000) 1173-1174. https://doi.org/10.1039/b001517m.

[98] K. Hara, K. Sayama, Y. Ohga, A. Shinpo, S. Suga, H. Arakawa, A coumarin-derivative dye sensitized nanocrystalline TiO_2 solar cell having a high solar-energy conversion efficiency up to 5.6%. Chem. Comm. 06 (2001) 569-570. https://doi.org/10.1039/b010058g

[99] K. Hara, T. Sato, R. Katoh, A. Furube, Y. Ohga, A. Shinpo, S. Suga, K. Sayama, H. Sugihara, H. Arakawa, Molecular design of coumarin dyes for efficient dye-sensitized solar cells. J. Phys. Chem. B.107 (2003) 597-606. https://doi.org/10.1021/jp026963x

[100] Z. Wu, T. Kinnunen, N. Evans, J. Yamagishi, C. Hanilçi, M. Sahidullah, A. Sizov. ASVspoof 2015: the first automatic speaker verification spoofing and countermeasures

challenge. in sixteenth annual conference of the international speech communication association. 16 (2015) 1-5. https://doi.org/10. 7488/ds/252

[101] M.S. Faber S. Jin, Earth-abundant inorganic electrocatalysts and their nanostructures for energy conversion applications. Energy Environ. Sci. 7(2014)3519-3542. https://doi.org/10.1039/ c4ee01760a

[102] J. Wu, Z. Lan, J. Lin, M. Huang, Y. Huang, L. Fan, G. Luo, Y. Lin, Y. Xie, Y. Wei, Counter electrodes in dye-sensitized solar cells. Chem. Soc. Rev. 46 (2017) 5975-6023. https://doi.org/10.1039/C6CS00752J

[103] J. Theerthagiri, A.R. Senthil, J. Madhavan, T. Maiyalagan, Recent progress in non platinum counter electrode materials for dye sensitized solar cells. Chem. Electro. Chem. 2 (2015) 928-945. https://doi.org/10.1002/celc.201402406.

[104] S. Yun, H. Pu, J. Chen, A. Hagfeldt, T. Ma, Enhanced performance of supported HfO_2 counter electrodes for redox couples used in dye sensitized solar cells. ChemSusChem. 7 (2014) 442-450. doi. 10.1002/cssc.201301140

[105] K. Kakiage, Y. Aoyama, T. Yano, K. Oya, J.-i. Fujisawa, M. Hanaya, Highly-efficient dye-sensitized solar cells with collaborative sensitization by silyl-anchor and carboxy-anchor dyes. Chem. Commun. 51 (2015)15894-15897. https://doi.org/10.1039/c5cc06759f

[106] S.Y. Myong, Recent progress in inorganic solar cells using quantum structures. Recent patents on nanotechnology. 01 (2007) 67-73. https://doi.org/10.2174/187221007779814763

[107] K. Barnham, P. Abbott, I. Ballsrd, D. Bushnell, A. Chatten, M. Mazzer, G. Hills, J. Roberts, M. Malik, P. O'Brien, Future applications of low dimensional structures in photovoltaics. Proc. Photovoltaics for the 21st Century,(2005) 30 - 45.

[108] R.Morf, Unexplored opportunities for nanostructures in photovoltaics. Physica E Low Dimens. Syst. Nanostruct. 14 (2002) 78-83. https://doi.org/10.1016/s1386-9477(02)00360-0

[109] C.B. Honsberg, A.M. Barnett, D. Kirkpatrick. Nanostructured solar cells for high efficiency photovoltaics. in 2006 IEEE 4th world conference on photovoltaic energy conference. 2006. https://doi.org/10.1109/WCPEC.2006.279769

[110] Bailey, S., S. Castro, R. Raffaelle, S. Fahey, T. Gennett, P. Tin. Nanostructured materials for solar cells. in 3rd World conference on photovoltaic energy conversion. 2003.2690-2693

[111] K. Zhao, Z. Pan, I.n. Mora-Seró, E. Cánovas, H. Wang, Y. Song, X. Gong, J. Wang, M. Bonn, J. Bisquert, Boosting power conversion efficiencies of quantum-dot-sensitized solar cells beyond 8% by recombination control. J. Am. Chem. Soc. 137 (2015) 5602-5609. https://doi.org/10.1021/jacs.5b01946

[112] R.K. Goyal, Nanomaterials and nanocomposites: synthesis, properties, characterization techniques, and applications: CRC Press, 2017.

[113] A. Kojima, K. Teshima, Y. Shirai, T. Miyasaka, Organometal halide perovskites as visible-light sensitizers for photovoltaic cells. J. Am. Chem. Soc. 131 (2009) 6050-6051. https://doi.org/10.1021/ja809598r

[114] J.H. Im, C.R. Lee, J.W. Lee, S.W. Park, N.G. Park, 6.5% efficient perovskite quantum-dot-sensitized solar cell. Nanoscale. 3 (2011) 4088-4093 https://doi.org/10.1039/C 1NR10867K

[115] C.W. Chen, H.W. Kang, S.Y. Hsiao, P.F. Yang, K.M. Chiang, H.W. Lin, Efficient and uniform planar-type perovskite solar cells by simple sequential vacuum deposition. Advanced Materials, 26 (2014) 6647-6652. https://doi.org/10.1002/adma.201402461

[116] M. Bag, Z. Jiang, L.A. Renna, S.P. Jeong, V.M. Rotello, D. Venkataraman, Rapid combinatorial screening of inkjet-printed alkyl-ammonium cations in perovskite solar cells. Mater Lett. 164 (2016) 472-475. https://doi.org/10.1016/j.matlet.2015.11.058

[117] J. Calbo, Dye sensitized solar cells: past, present and future. photoenergy and thin film materials, in: X Yu Yang (Eds) Photoenergy and thin film materials, Scrivener Publishing LLC. (2019) 49-119.

Materials for Solar Cell Technologies II
Materials Research Foundations **104** (2021) 24-39

Materials Research Forum LLC
https://doi.org/10.21741/9781644901410-2

Chapter 2

Versatile Applications of Solar Cells

Vivian Chimezie Akubude [1], Ugochukwu Kingsley Ebisike[2], Kevin Nnanye Nwaigwe[3],
Jelili Aremu Oyedokun[4], Ayooluwa Peter Adeagbo[5]

[1]Department of Agricultural and Bioresource Engineering, Federal University of Technology, Owerri, Imo State, Nigeria

[2]Department of Electrical and Electronics Engineering, Federal University of Technology, Owerri, Imo State, Nigeria

[3] Department of Mechanical Engineering, University of Botswana, Botswana

[4] Engineering and Scientific Services Department, National Centre for Agricultural Mechanization, Ilorin, Kwara State, Nigeria.

[5]Department of Electrical and Electronics Engineering, Adeleke University, Ede, Osun State, Nigeria

akubudevivianc@gmail.com

Abstract

Solar cells have changed the way electricity is generated; it helps the world to reduce carbon emission, and consequently makes our electric grid system more resilient and reliable. Hence, this chapter presents the concept of solar cells and the basic principle of operation. The chapter also discusses materials in construction of solar cells including conventional semiconductors such as silicon and emerging/organic materials such as perovskite and quantum dots. Various applications of solar cells which include space research, telecommunications, grid connections, stand-alone connections and off-grid applications are also highlighted. Given the versatile application of solar cells, it is the future of electricity generation.

Keywords

Solar Cells, Photovoltaic Cells, Energy, Applications, Silicon

Contents

1.   Introduction

Solar cells are devices that convert light energy into electrical energy [1]. Solar cells have found applications in several devices including calculators and satellites. Its earliest use was majorly in space but presently they are in more common applications. During the last century, the world has seen phenomenal changes in the power industry due to science and technology. There are different methods of generating electrical power used in powering up our electrical appliances. This chapter explores the use of solar cells for generating power using energy derived from the sun and its versatile applications.

Solar cells or photovoltaic cells as it is sometimes called converts solar energy into electricity via the use of solar panels [2]. It traps energy from the sun converting it into electricity. It is made of bluish black colouration and usually octagonal in shape. Solar cells can be thought of like cells in a battery. They are bundled into larger units to form solar panels. A beam of sunlight actually contains smaller particles known as photons, when a solar cell comes between the paths of this beam of sunlight, the photons are intercepted by the solar cell which converts them into free-flowing electrons, and these free-flowing electrons are what constitute electric current. Each cell produces a few volts of electricity (about 0.5 volts for silicon cell), to get a specific voltage level that will be suitable for use; a reasonable number of solar cells must be bundled together into modules, panels and array. They possess certain qualities that make them stand out among other energy sources, some of these qualities include durability, portability, low maintenance requirement and they are also used to generate power in remote conditions, i.e., they can be deployed for powering electrical devices from the closest electrical outlet [3].

Solar technique of power generation is incredibly amazing, clean, very easy and cheap to implement, this has prompted lots of research and exploration into how to turn it into a stand-alone power generation scheme for both small and large machines and buildings. With solar, the need to spend time, money and effort to drill oil, mine coal or do some other form of strenuous activities that involved tilling the earth to get resources most of which needs to be used under controlled conditions in order to get power are eliminated. The sun itself is a natural mega power station, capable of supplying the world with enormous amount of energy for as may be required or demanded and the biggest advantage of it is that it never goes off. The sun produces a carbon-free source of energy.

There are also disadvantages of solar energy, but they are very few and currently being researched to be improved upon in usage. Energy variation is a prominent challenge of solar energy. The energy level trapped from sunlight is at its lowest point at morning and evening respectively while it is at its peak at midday when the sun is directly overhead. Another common challenge is production and processing equipment. Production of solar panels and other devices used to harness the sun's radiation and transform it to electrical energy in suitable voltages for transmission and consumption tends to involve high level technical expertise and research together with funds. These are required to produce good solar panels that can effectively transform solar energy to electrical energy. Solar energy via solar cell is converted directly into direct current (DC), and since most electrical devices or equipment run on alternating current, an inverter is needed to convert this direct current (DC) supply by the solar panels into AC in a suitable voltage that will be viable for transmission and consumption by home appliances.

Finally, engineers will need to design new devices which are easily powered directly from the output of solar panels. Bad weather conditions and dark hours also contribute to the challenges encountered in solar technology. One of the effects of revolution of the earth is that it causes changes in the sun's visibility from the earth and this in turn gives rise to changes in weather conditions which in turn brings about seasons. In tropical Africa, there are basically two seasons: the wet (rainy) season and the dry season while four seasons exist in Europe, America and some parts of Asia, they are namely: autumn,

Materials Research Forum LLC
https://doi.org/10.21741/9781644901410-2

spring, summer and winter. The summer season in the western world is similar to the dry season in tropical Africa and this season is known for minimal or no rainfall and maximum sunlight. The solar energy gotten during this period is maximum and this makes it the peak season for using solar technique for power generation. Other seasons are characterized by rainfall and poor visibility (though not all the time). The solar energy harnessed during this time is minimal. An effect of rotation of the earth is that it brings about day and night. Solar activity during the day is high while solar activity at night is not even present. This means that at night, devices being powered directly via solar panels will not function because their source of energy, the sun, is not present. This is a major drawback because it shows that solar technology cannot work at certain periods. However, with the discovery of secondary cells (cells that can be recharged), this drawback is completely eliminated, because the energy from the sun could be stored in these batteries and used during the dark periods or periods with weather conditions that causes low visibilities. Similarly, cost of construction of storage facilities is also a concern. It costs a lot to construct large batteries which are used to store the converted electrical energy from the sun by the solar panels. It also takes a lot of batteries to store electrical energy that can be used to power an entire factory [4]

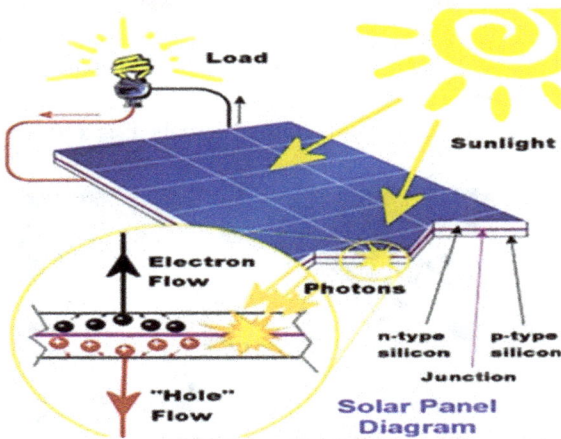

Figure 1 basic principle of operation of solar cell [4]

Materials for Solar Cell Technologies II   Materials Research Forum LLC
Materials Research Foundations **104** (2021) 24-39   https://doi.org/10.21741/9781644901410-2

2.   Working principle of solar cell

Solar cells operate on the principle of photovoltaic effect which literally means voltage produced from light. Figure 1 [5] describes the fundamental principle of operation of solar cell.   It comprises of a sandwich of n-type semiconductor and p-type semiconductor. Photons from sunlight hit the upper surface of the cell, the photons which are depicted by the yellow arrows and star, carry their energy downwards through the cell. Also, they donate their energy to electrons in the p-type layer. This energy is utilized by the electrons to overcome the hindrance at the layer by moving over it up to the n-type layer and finally it gets out into the circuit. The excess flowing electrons through the circuit then lights up the bulb.

3.   Applications of solar cells

3.1   Space research

Solar cells are utilized in powering space vehicles, satellites and telescopes, because they tend to supply the world with certain information about the earth and they are also used to relay telecommunication signals, going off abruptly due to diminishing fuel sources could be catastrophic for people who rely on them for certain information. Also, it cost much travelling to space to refuel these machines and vehicles. The only way of mitigating and consequently eliminating the need for refuelling by travelling back to space is by harnessing the power of the sun and converting it into electrical energy that will be needed to power these vehicles. Using solar cells to power space vehicles is very economical and reliable as the cost incurred is only the cost of manufacturing and attaching solar.

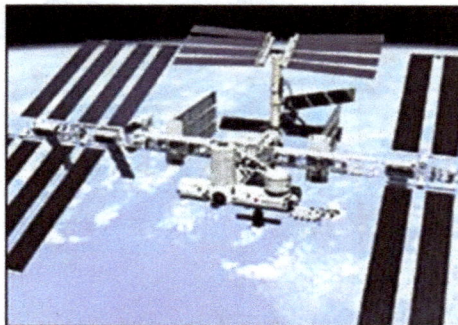

Figure 2 International space station [6]

The Figure 2 [6] shows the international space station which is still under construction. It is an example of a solar powered station. It is proposed to be the most powerful solar array in space containing 250,000 solar cells and the entire array will be able to power a small neighbourhood. Also, Figure 3 [7] shows the picture of a satellite in space. Attached to its sides are solar panels to power it up by converting the energy of the sun into electrical energy.

Figure 3 Picture of a satellite in space [7]

3.2   Telecommunications

In our world today, we desire to have an efficient communication system that enables us to reach any part of the world we want to. In order to achieve this, we have to deploy telecoms infrastructure in certain remote and isolated areas which ranges from hills and mountain tops to deserts and forest regions. The effect of this will tell on both cost of installation, operation and maintenance on the particular firm. It is a known fact that the aim of every business is to make profits and to achieve this in this context, the operator must establish self-sustainable mobile networks with higher efficiency and profitability and remains competitive in a lower Average-Revenue-Per-User (ARPU) environment. The best way to make this a reality is by deploying solar panels and batteries in such remote locations where these Base Transciever Station sites will be constructed so as to be the major source of power supply to the site. Recall that the cost of construction and deployment of solar panels to power these telecom equipment is nothing compared to outages and cost of transportation to access these sites regularly, also the sun only goes down at night and the energy converted during the day is already stored in batteries to compensate for the unavailability of the sun at night, the change-over process will be made automatic via a contactor so as to prevent outage and manual handling. Furthermore, Signals required by communication systems need amplification after certain intervals so as not to degrade. Various relay towers are stationed to boost high grounds which are mostly favoured as the sites for repeater stations. In remote applications, these stations are far from the electric grid and may not be served by electricity from the grid, in such scenario, there will be breach in communication, because it will be tasking and costly to set up power stations at remote locations to power up repeaters. To reduce the difficulty, solar devices and systems are being installed as a viable alternative [8]

3.3   Grid connected applications

When using solar panels in a community's grid system, the modules are connected to an inverter such that the output voltage from the panel which is in direct current (DC) will be converted to alternating current (AC) voltage which is used by our domestic and industrial appliances [9]. Leftovers from this system can then be sold to the electric grid of a community from which it could be used to power up other homes under the control of the grid as shown in Figure 4 [10]. However, there is a limitation to the use of solar cells in grid systems. The constraint is due to the fact that solar cells only operate when there is sunlight. In night hours and periods of bad weather where visibility is poor, they do not work and as such a backup system, usually a battery is required to make sure that the electricity supply is not interrupted. Systems of this nature are most commonly used in houses or commercial buildings to offset electricity cost. A well-designed solar cell electricity generating system with a proper storage facility can be a very intelligent method of displacing electricity usage from the grid during peak hours [8].

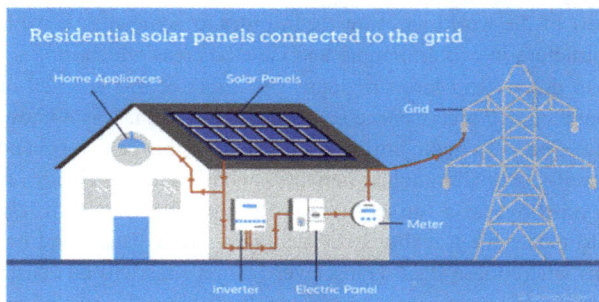

Figure 4 Grid connected residential solar pv energy [6,10]

3.4   Stand-alone devices

With the recent advancement in solar technology, certain systems are now being digitized in such a way that they can make use of the direct current (DC) voltage generated by the solar panels at the particular magnitude, and they do not rely on or require the electric grid system. These devices are sometimes referred to as stand-alone devices. In the night or in the event of bad weather conditions where sunlight visibility is poor, a battery storage system is used. There are lots of devices that directly use solar energy for its operations and they use batteries to ensure continuity of power during the night and bad weather condition. Some of them are listed and briefly explained below:

Solar water heaters: they are basically water heaters made up of collectors and storage tanks as shown in Figure 5 [11]. These systems use the Sun to heat water in collectors generally mounted on a roof. The heated water is then stored in a tank similar to a conventional gas or electric water tank [11].

Fiure 5 Solar heater [11]

Solar street light: in our world today most street lights found on the streets are being powered by solar technology as opposed to the conventional way of powering them up via the grid system. They operate of the same principle of solar cells, just that this time, their output which is DC is directly used to power up a DC light. The solar street lights comprise of solar cells, which absorb the solar energy during daytime. The photovoltaic cells convert solar energy into electrical energy, which is stored in the battery. The lamp has already been programmed by a microprocessor to come on after certain hours, and go off once it senses sunlight, and it consumes the electricity already stored in the battery. During the day- time the battery gets recharged and the process keeps on repeating every day. Figure 6 [12] shows the picture of a solar street light.

Figure 6 Solar street light [12]

Portable lightning and charging systems:

Nowadays, torch lights and certain LED lamps and other home appliances that makes use of rechargeable batteries like fans, radios, refrigerators, mobile phones and power banks are now designed in such a way that their batteries are being charged directly from the sun via solar cells as shown in Figure 7a-e [13]. This technique makes these devices portable and free from charging by electricity supplied through the national grid, thus saving energy cost for the user and making the device more mobile [8]

Solar powered pumping machine:

Solar powered pumping is another exciting device that makes use of renewable energy by way of harnessing the energy from the sun to drive a motor which drills water from an open well, stream or underneath the ground. The system basically uses solar cells in a solar photovoltaic system to convert the energy from sun into electrical energy which is used for the smooth running of the motor pump. This system makes it easier to use the pump anytime there is sunlight and it eliminates the need to power the pumps via electricity supplied through the national grid. This system finds great applications in countries with low power supply and in countries where clean water supply is poor and not centralized. It also finds applications in agriculture as it can be used for irrigation purposes. It basically consists of a solar PV system with a surface mounted centrifugal pump which is compatible with the solar panel (a direct current motor) and the series of pipes to channel the water pressure [14].

a. solar powered reading lamp

b. solar powered keyboard

c. solar powered window charger

d. solar powered sound system

e. solar powered flashlight

Figure 7a – e Portable lightning and charging systems [13]

3.5   Solar photovoltaic winnower cum dryer

This is a suitable device for separating threshed agricultural produce and also for drying fruits and vegetables with forced circulation of air. The system as shown in Figure 8 [14] is made up of PV module, compatible winnower, pre-air heating tunnel and specially designed solar drying cabinet. Research shows that the system is more efficient when compared with open sun drying. This is also seen when different fruits and vegetables gets dried in less than half the time it will take to dry under direct sun. The overall quality of the dried material is improved by improving the fan speed with more irradiance. The system can be enhanced by providing certain properties for illumination coupled with the addition of a battery and charge regulator, therefore making it a suitable device for processing different agricultural produce and utilizing the generated PV electricity for one or another purpose throughout the year [14].

Figure 8 Solar PV winnower cum dryer cum [14]

3.6   Solar photovoltaic duster

Crop safety and security against pest and diseases is an essential aspect of agriculture. Solar PV duster is a new device used for spraying insecticide and pesticide powder on crops. It basically consists of a small unit solar PV panel carrier, battery compatible dusting unit. The panel can concurrently charge the battery and provides shade for the user (as shown in Figure 9) [12]. The system can run an ultra-low volume (ULV) sprayer and be used to light up a light emitting diode (LED). The device can be used for lighting the house throughout the year and for operating the duster/sprayer as and when required [12].

Fig. 9 solar PV sprayer [14]

3.7   Solar powered cars

With the world delving into use of renewable energy systems to produce electricity in order to reduce carbon emissions, the need to manufacture cars that will run on renewable energy is very crucial. Solar powered cars are vehicles that use electrical energy resulting from converted sunlight energy to drive the car [15]. This electrical energy serves as fuel in driving the car. They have the capacity to store solar energy and utilize at night or in the absence of direct sunlight. Its use is advantageous in terms of environmental and noise pollution reduces fuel costs, sustainable and eco-friendly and doesn't incur additional costs except battery replacement.

However, when designing photovoltaic vehicles, there are two drawbacks, namely: aesthetics of the car, which must accommodate the solar panels in terms of weight and beauty and also the cost of mining these solar cells. Research shows that the present marketable operated solar panels that are in use have just 15 to 20% efficiency; this implies more solar panels will be required to power a solar car. But achieving this will create two problems – weight and cost.

3.8   Solar aviation

In order to meet the demands of achieving a friendly ecosystem characterized by low carbon emission into the atmosphere, the aviation industry which runs mainly on fossil fuels (crude oil) must be made to run entirely on renewable energy sources and one of the best and efficient way of achieving this is by the use of solar cells on all aircrafts, the outcome of this will be known as solar powered aircraft. The principle of operation of a solar powered aircraft is very much like that of the conventional aeroplane only that it doesn't make use of fossil fuels and it emits no CO_2 as such. In a solar powered aeroplane for example, certain parts of the wings are lined up with solar cells which are usually bundled to form solar panels. During a flight, the operation of a solar powered plane automatically switches between battery and solar power. The solar cells are used to convert solar energy from the sun into electrical energy; this electrical energy is then used to power electrical propellers which in turn drive the flight of the plane. When the sun is active, solar energy which is converted into electrical energy runs the aircraft's propellers and charges the batteries or fuel cells used by the plane to its peak. In order to achieve faster charging of the batteries to its peak, the pilot needs to reduce his flight cruising speed. During the dark periods (at night or in clouds), the batteries or fuel cells control the operation of the flight propellers.

However, solar planes are not the future of flight; this is due to certain reasons some of which includes:

- Inability to carry enough passengers and cargoes because the solar system and the storage battery have already taken most space contributing to weight of the plane.

- Low energy density; the weight of lithium-polymer batteries is a bit much and when loaded on to the plane, it adds more weight to the plane and this greatly increases the energy needed for take-off. Except research is done to produce light weight solar panels and batteries, it may be unlikely to have solar planes as part of the flying vehicles on earth. [16,17]

3.9   Off-grid power

Solar cells also find wide application as an off-grid energy source. For instance, several traffic, emergency and construction road signs and also electronic bill boards use solar cells for power during the day time. Such equipment are also loaded with batteries to power them up at night, decreasing the need for gasoline-powered generators for remote and mobile uses. Also some scientific research centres are situated far away from the town especially those used for weather and mining purposes, these facilities needs power to facilitate their activities. Since they are situated far from the heart of the town, the cost of extending transmission/distribution line to such location is capital-intensive coupled with high power loss due to the distance involved. As a result, solar PV system is used to power these facilities. Moreover, solar panels can be used to provide electricity for rural areas that are far away from the town and not connected to the grid. Solar cells bundled together to form solar panels could be used to build a charging station for electric vehicles. Finally, in areas where natural disasters occurs, solar panels are used to provide temporary power supply to such areas so as to speed up evacuation of people and valuables from the area, and carrying out restructuring and maintenance of the location [18].

3.10 Medical applications

Solar cells also find applications in medicine; they can actually be placed on the skin to continuously recharge implanted electronic medical devices. Most implanted devices are electronic in nature and their power is supplied via batteries and these batteries in turn greatly influence the sizes of these devices. To enable these devices last long for their purposes, more batteries are added and some are actually redundant so that when one stops, the other takes over, but this in-turn creates a larger device. But the challenge is that when the power in such medical device batteries runs out, it will necessitate either recharging or replacement. In most cases this means that patients have to undergo implant replacement procedures, which is not only expensive and tedious but also holds the risk of medical complications. Current research conducted in Switzerland shows that about 12 microwatts could be generated by 3.6 centimetres square of solar panel which is sufficient enough to power up or extend the battery life of a pacemaker or any other active human implant device [19].

3.11 Solar powered inverters and uninterruptible power supply (UPS)

An inverter is basically a device that converts direct current usually supplied by a battery into alternating current while Uninterruptible Power Supply (UPS) is a device that serves as battery backup when the electrical power fails or drops to low voltage level. In mission critical data centres, UPS systems are used for just a few minutes until electrical generators take over. UPS systems can be set up to alert file servers to shut down in an orderly manner when an outage has occurred, and the batteries are running out. The inverter and the UPS can be combined together to form online uninterruptible power supply (UPS). In online uninterruptible power supply (UPS) system, the inverter perpetually provides clean power from the battery while it charges it, and the computer and other sensitive equipment is never receiving power directly from the AC outlet. Nevertheless, online UPS units contain noiseless cooling fans which may need location planning for home user or small office. The inverter discussed here is charged directly from an AC source usually gotten from the central electric grid. However, with solar inverters, there is virtually no need for AC charging except at night periods when the sun is not present. The sun will continuously charge the battery and provide DC power for the inverter to use while the UPS makes sure the devices attached to it never go short of power in the exact voltage required by them [20].

3.12 Hybrid applications

Solar PV technology that comprises of solar cells can be used with other power generating systems using both renewable and non-renewable energy sources. This is referred to as hybrid. This is done in order to have a higher power delivery to the load system and also at a lower cost than when only PV technology is used if properly controlled. Examples include PV-wind hybrid, PV-wind-diesel hybrid, PV-diesel generator etc.

Conclusions

The use of solar cells is still developing, more and more applications keep coming and with time, a good percentage of the world's energy will be generated using solar cells. Solar cells are a good substitute to global air pollution emanating from energy generation using fossil fuels and they can be made portable with little cost and efforts when compared to what is involved in drilling oil and mining coal. This chapter gradually explained solar cell and its applications. It is important to support solar technology so as to reduce global warming and all other hazards associated with the use of fossil fuel and also it is cheaper when deployed to use in some respects.

References

[1] NREL. 'An overview of the materials used for solar cells'. Retrieved from https://www.azooptics.com/Article.aspx?ArticleID=147 on 24 December 2019.

[2] Suresh S. Recent trends on nanostructures based solar energy applications: a review. Rev. Adv. Mater. Sci. 34 (2013) 44-61.

[3] Main uses of solar cells. Retrieved from https://www.education.seatlepi.com/main-uses-solar-cells-3343.html 2013.

[4] Solar cells. Retrieved from https://www.explainthatstuff.com/solar cells.html, 2019.

[5] Working principle and development of solar cell by kynix semiconductor. Retrieved from https://www.apogeeweb.net, 2017.

[6] Uses of solar cells. Retrieved from https://www.chm.bris.ac.uk/webprojects 2003/ledlie/uses_of_solar_cells.htm, 2015.

[7] Sambasiva RaO, V. Extend LEO downlinks with GEO satellites. Retrieved from https://www.mwrf.com/systems/extend_leo_downlinks_geo_satellites, 2016.
[8]Satheesh, K. Applications of photovoltaic systems. Retrieved from https://www.labri.fr

[9] K.N. Nwaigwe, P. Mutabilwa and E. Dintwa. An overview of solar power (PV systems) integration into electricity grids. Materials Science for Energy Technologies, 2(2019)629–633. https://doi.org/10.1016/j.mset. 2019.07.002

[10] Luke T. How to choose the best solar panel for you. Retrieved from https://news.energyusage.com/how_to_choose_the_best_solar_panel_for_you/, 2017

[11] Solar devices. Solar water heaters. Retrieved from https://www.solardevices.net

[12] RECW. About solar street light. Retrieved from https://recw.in/SOLAR-STREET-LIGHT.html

[13] Damn, A. 12 awesome gadgets that every home should be using. Retrieved from https://www.makeuseof.com/tag/12_awesome_solar_powered_gadgets_every_home_using/, 2015.

[14] Basics of solar water pumping system. Retrieved from https://vikaspedia.in/energy/energy_production/solar_energy/solar_water_pumping_system

[15] Kashyap, V. All you need to know about solar-powered cars. Retrieved from https://interestingengineering.com/all-you-need-to-know-about-solar-powered-cars, 2019.

[16] Solar Impulse Plane Takes Flight. Retrieved from https://solarearthusa.com/solar-impulse-plane-takes-flight/, 2019.

[17] Brad, P. Solar planes aren't the green future of air travel. But here's what could be. Retrieved from https://www.vox.com/2016/5/6/11569202/aviation-emissions-solar-plane, 2016.

[18] The beating heart of solar energy. Retrieved from https://www.Springer.com/gp/about-springer/media/research-news/all-english-research-news/the-beating-heart-of-solar-energy/11952482, 2017.

[19] Definition of: UPS.. Retrieved from https://www.pcmag.com/encyclopedia/term/53509/ups

[20] Uninterruptible Power Supply Circuit Diagram and working. Retrieved from https://www.elprocus.com/types-of-uninterruptible-power-supply-devices

Materials for Solar Cell Technologies II
Materials Research Foundations **104** (2021) 40-76

Materials Research Forum LLC
https://doi.org/10.21741/9781644901410-3

Chapter 3

Applications of PEDOT:PSS in Solar Cells

Gokul Ram Nishad[1], Younus Raza Beg[1], Priyanka Singh[1*]

[1]Department of Chemistry, Govt. Digvijay PG Autonomous College, Rajnandgaon-491441, Chhattisgarh, India

*priyankasingh121@yahoo.com

Abstract

Poly(3,4-ethylenedioxythiophene)-poly(styrenesulfonate) (PEDOT:PSS) is increasingly being used in the field of printed and flexible electronics in the form of electrode as well as intermediate layer. PEDOT:PSS belongs to the family of intrinsically conducting polymer materials whose members can conduct electricity in spite of their organic nature without the presence of metals. It is non-toxic, stable in the presence of air and humidity. Above all, it can be easily processed through conventional means. This chapter deals with the applications of PEDOT:PSS in organic solar cells (OSCs), dye sensitized solar cells (DSSCs) and silicon based hybrid solar cells. PEDOT:PSS is being used as electrode, buffer layer and hole conductive layer. It could manipulate the catalytic nature of counter electrode used in DSSCs. Whereas it may help to manipulate the morphological character in Si based hybrid solar cells along with enhancement of cell performance.

Keywords

Dye Sensitized Solar Cells, Hole Transport Layer, Inverted Organic Solar Cells, Organic Solar Cells, PEDOT:PSS, Silicon Based Hybrid Solar Cells

Contents

1. Introduction

The solar or photovoltaic cells (SCs), converting light into electrical energy is grouped into three generations. The traditional first generation solar cells are composed of crystalline silicon. The second generation thin film solar cells are composed of amorphous silicon, cadmium telluride (CdTe) and copper indium gallium selenide (CIGS) cells. The third generation thin-film solar cells are known as emerging photovoltaics because they are under exploration stage and can be based on inorganic or organic materials, organometallic compounds, composites, polymers, etc. Inorganic materials are being replaced with conjugated polymers in solar cells due to several advantages such as cheap processing, eco-friendly technology, etc. Out of these, conducting polymers come under the prime focus of scientific community in order to attain cheap and highly efficient solar cells.

Poly(3,4-ethylenedioxythiophene):poly(styrenesulfonate) (PEDOT:PSS) (Scheme 1), the intrinsically conducting polymer, is widely used in doped state for preparing solar cells, supercapacitors, sensors, organic light emitting diodes (OLEDs), etc. due to its high electrical conductivity. The PEDOT:PSS film is highly transparent in the visible region, extremely flexible and thermally stable. Therefore, it can be used as organic solar cells (OSCs) anode buffer layer. Low-resistivity PEDOT:PSS layers increase the performance of OSCs by improving the power conversion efficiency (PCE). Platinum (Pt) exhibits efficient catalytic activity for triiodide/iodide redox reaction but it is expensive. PEDOT:PSS is an option to develop cheaper counter electrode (CE) replacement of this expensive metal in dye sensitized solar cells (DSSCs).

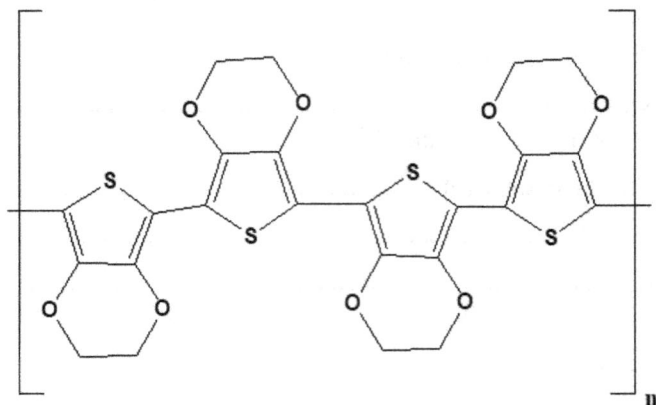

(a) **PEDOT**

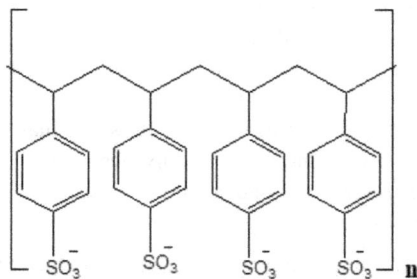

(b) **PSS**

Scheme 1 (a) Structure of PEDOT (b) structure of PSS.

PEDOT:PSS may serve as the catalyst for triiodide/iodide redox reaction. Its excellent catalytic activity and low cost has attracted researchers from around the world. PEDOT:PSS acts as hole transport layer (HTL) and helps in interface improvement along with gap reduction between active layers and indium-tin oxide (ITO) lowest unoccupied molecular orbital (LUMO) levels. The phase separation between PEDOT and PSS causes low conductivity and improper contact between PEDOT:PSS and organic layer [1]. Surface morphology and conductivity of PEDOT:PSS films depends on particle size, PSS proportion, viscosity of solution, etc. This chapter deals with application of PEDOT:PSS in OSCs, inverted OSCs, DSSCs and Si based hybrid solar cells.

2.   Organic solar cells

Organic solar cells (OSCs) also known as plastic solar cell are based on organic electronics and utilize small organic molecules or conductive organic polymers for charge transport and light absorption. They generally contain electron donor and acceptor layers sandwiched between electron/hole layer and electrode. Blocking layer and electrode used depend on the structure of device, whether regular or inverted. Only difference between the regular or inverted devices is that the direction of exit of electric charges is opposite in the two due to the reversal of positive and negative electrodes. Inverted OSCs offer longer lifetimes as compared to the regular OSCs.

2.1   PEDOT:PSS as electrode

Kawano et al. (2006) investigated polymer and fullerene derivative based stacked bulk hetero-junction (BHJ) OSCs obtained by spin-coating method. Poly(2-methoxy-5-3,7-dimethyloctyloxy)-1,4-phenylene vinylene (MDMO-PPV) : [6,6]-phenyl-C_{61}-butyric acid methyl ester (PCBM) blend and 40 nm thick PEDOT:PSS film were used as the was used as the active layer and hole conductive layer, respectively [2]. Open-circuit voltage (Voc) is independent of the number of photoactive layers. It is determined by the addition of Voc of the two stacked BHJ OSCs. Voc of 1.34 V was obtained which is about 1.6 times larger than that of single BHJ OSCs. Hoppe and co-workers [3] determined the optical constants for polymer/fullerene based BHJ OSCs having a structure of glass\ITO\PEDOT:PSS\(3,7-dimethyloctyloxy methyloxy poly(paraphenylene vinylene):(1-(3-methoxycarbonyl)propyl-1-phenyl [6,6]C61) blend [MDMO-PPV:PCBM]\Al. Model dielectric functions were fitted to reflection and transmission spectra of the layers present in the OSCs.

Transparent gold (Au), copper (Cu), and silver (Ag) metal-wire electrodes were prepared through nanoimprint lithography on glass substrate coated with PEDOT:PSS [4]. Poly(dimethylsiloxane) (PDMS) was used as a mold. PDMS stamp was prepared through a nano imprinted template and 40 nm-thick layer of Cu was deposited on it. The PEDOT:PSS film was spin-casted on a glass substrate and PDMS stamp was coated over it. Patterned Cu electrodes were obtained on the PEDOT:PSS layer after peeling the PDMS stamp. Anode's work function for Au, Cu, and Ag were found to be 5.22, 4.65 and 4.0 eV respectively. Metal electrodes having work function lower than PEDOT:PSS offered better charge transfer to the electrode from the PEDOT:PSS layer. This resulted into enhanced fill factors (FFs) and power conversion efficiencies (PCEs). The optical transmittance of Au, Cu, and Ag based electrodes along with indium tin oxide (ITO) based control device measured in visible region with reference to air. ITO, Au, Cu, and Ag electrodes have transmittance of 87, 84, 83 and 78 %, respectively in the visible range. Ag electrode has transmittance lower than Au and Cu electrodes in spite of similar line-width and thickness due to dispersion properties of Ag. ITO, Au, Cu, and Ag electrodes had sheet resistance of 12, 24, 28, 23 $\Omega.sq^{-1}$ respectively. Decrease in line-width is responsible for higher sheet resistance. Thicker metal layer deposition can compensate this issue. The current-voltage properties of Au, Cu, and Ag electrode based solar cells and were found similar to the ITO electrode, demonstrating the interchangeability of these electrodes.

Kang and co-workers (2010) [5] have reported a transparent electrode of Cu nanowire mesh which was prepared through transfer printing using the flexible PDMS stamp (Figure 1) [5]. The Cu/Ti layer was given a short O_2 plasma treatment followed by transferring the metal to 30 nm thick PEDOT:PSS layer which was spin coated over PET substrate. This electrode exhibits high flexibility, optical transmittance and conductance as compared to the common ITO electrode settled on plastic substrates.

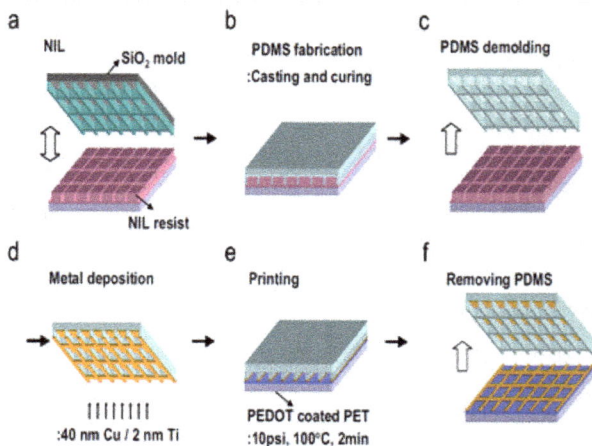

Figure 1 Schematic diagram of PDMS stamp fabrication and the Cu nanowire mesh electrode transfer printing (Reprinted with permission from Elsevier, [5]).

Materials for Solar Cell Technologies II   Materials Research Forum LLC
Materials Research Foundations **104** (2021) 40-76   https://doi.org/10.21741/9781644901410-3

Ultrathin and light weight OSCs with improved flexibility offering 4% PCE were developed by Kaltenbrunner et al. (2012) [6]. These devices were capable of surviving a quasi-linear compression of less than 70%, cyclic stretching and compression of 50% for about 20 full cycles without much decrease in device performance. Qu et al. (2011) [7] studied the role of plasmonic metal nanoparticles in improvement of optical absorption of thin film Poly(3-hexylthiophene) : phenyl-C61-butyric acid methyl ester (P3HT:PCBM) OSCs through finite element method and 3-D model. When Ag nanoparticles were used at active layer and anode layer (PEDOT:PSS) interface, plasmonic enhancement of more than 100% could be gained.

Kymakis et al. (2007) [8] prepared single wall carbon nanotube (SWCNTs) and PEDOT-PSS blends in order to be used as hole collecting electrode in P3HT:PCBM (donor:acceptor) OSCs. This improved the conductivity and optical transparency. Devices having PEDOT:PSS-SWNTs/P3HT-PCBM/Al, ITO/PEDOT:PSS/P3HT-PCBM/Al and PEDOT:PSS/P3HT-PCBM/Al were studied. PCE of 1.3%, FF of 0.4, Voc of 0.6 V and short-circuit density (Jsc) of 5.6 mA/cm^2 were obtained for the former device under white light illumination (100 mW/cm^2). The results are much similar to that of the ITO glass substrates based reference cells having identical structure and fabrication. However, the fill factor is very low because of the high resistance of the PEDOT:PSS and SWCNT film. But, this work provides a low cost substitute for ITO. The current-voltage (I-V) curves show that introduction of SWCNTs to PEDOT:PSS improves the device performance because the former improve the conductivity of the later. PEDOT:PSS-SWCNTs electrode based devices provide, Jsc slightly lower and Voc similar to the ITO based reference cell. PEDOT/PSS-SWCNTs have a sheet resistance twenty higher and a much lower PCE than that of ITO based devices. Low fill factor is due to the high series resistance of the PEDOT/PSS-SWCNTs.

Li et al. (2015) [9] studied the effect of ethylene glycol additive, sulfuric acid post-treatment and polyethylenimine (PEI) treatment in P3HT: indene-C60 bisadduct (ICBA) based OSCs. PEI treatment reduces the PEDOT:PSS film conductivity. 2-methoxyethanol processing solvent improves the conductivity whereas isopropanol or water renders the conductivity unaltered for pristine PH1000 film. Device with glass/low-WF PEDOT:PSS/P3HT:ICBA/high-WF PEDOT:PSS structure provide a Voc of 0.82V, FF of about 0.62 under white light illumination (100 mW/cm^2). Park et al. (2010) [10] used transparent electrodes of graphene sheets developed by chemical vapor deposition in copper pthalocyanine and fullerene (CuPc\C$_{60}$)based OSCs. PCE similar to ITO electrodes was obtained. But surface wettability between the electrodes and hole-transporting layer remains a major challenge for pristine graphene electrodes due to its hydrophobic nature. Doping of gold (III) chloride (AuCl$_3$) in graphene changes the

Materials for Solar Cell Technologies II   Materials Research Forum LLC
Materials Research Foundations **104** (2021) 40-76   https://doi.org/10.21741/9781644901410-3

surface wettability leading to formation of uniform hole-transporting layer and improved device success. Doping also enhances the conductivity and PCE of the device.

Noh and co-workers (2014) [11] fabricated ITO-free OSCs based on composite electrode of Ag nanowire (AgNW) and PEDOT:PSS. The electrode was prepared by single step spray coating of AgNW and dimethyl sulfoxide (DMSO)-treated PEDOT:PSS mixture. Spray deposition time was the prime factor for controlling the optical transmittance, thickness and sheet resistance of the electrode film. PCE of about 2.16% was obtained under $100 mW.cm^{-2}$ and AM1.5G illumination. When an extra spray-coated buffer layer of PEDOT:PSS was used, it led to smoothening of the composite electrode surface, decreased shorting and increased cell-efficiency about 2.65%, similar to ITO-based devices. DC conductivity study of sorbitol doped PEDOT:PSS films by Park and co-workers (2011) [12] shows that work function decreases with increase in the concentration of dopant. This was also confirmed by UV photoelectron spectroscopic studies. ITO and PEDOT:PSS have ohmic contact and PEDOT:PSS work function determines the pertinent hole extraction barrier and affects the performance of cells. A preliminary PCE of 3.26% was obtained. Singh and co-workers (2015) [13] have reported the ink-jet printing technique for obtaining transparent and smooth PEDOT:PSS films for using in OSCs. Surface treatment and drop spacing along with substrate temperature throughout the printing and annealing process are the major parameters affecting continuous film formation. Transmittance of 90% was obtained for a 110 nm thick film which was produced by using 30 mm drop spacing and 120°C annealing temperature. The OSCs based on inkjet printed PEDOT:PSS films had a PCE of 2.64% as compared to 3.1%, for OSCs based on common spin coated PEDOT:PSS films.

2.2   PEDOT:PSS as hole transport layer

PEDOT:PSS hole transport layer (HTL) helps in efficient hole transportation, interface improvement and gap reduction between active layers and ITO LUMO levels. Generally PEDOT:PSS layer is spin coated which does not suit large area coating. Low-resistivity PEDOT:PSS layers increase the PCE of OSCs. Therefore, while using PEDOT:PSS, it becomes important to reduce its resistivity in order to stop the movement of generated holes towards the metal electrode. Bulk heterojunction (BHJ) based OSCs offer several advantages including cost effectiveness, eco-friendliness and easy production. They find a lot of applications because poly(3-hexylthiophene) (P3HT) and (6.6) phenyl-C61-butyric acid methyl ester (PCBM) (electron donor and acceptor) based OSCs exhibit PCEs more than 3% (AM 1.5 G illumination). Manipulation of donor and acceptor, electrode and buffer layer as well as annealing temperature can improve the performance of the devices.

BHJ OSC based on 1,4,8,11,15,18,22,25-octahexylphthalocyanine (C$_6$PcH$_2$) and PCBM were prepared through spin-coating. PEDOT:PSS served as a hole transport layer and was spin-coated over ITO coated quartz substrate. A PCE of 3.1% and external quantum efficiency higher than 70% in Q-band region of C$_6$PcH$_2$ were obtained. [14]. Metal chlorides like lithium chloride (LiCl), sodium chloride (NaCl), cadmium (II) chloride (CdCl$_2$) and copper (II) chloride (CuCl$_2$) were added to PEDOT:PSS hole transport layer in P3HT:PCBM BHJ OSCs and their effect on conductance, transmittance and morphology was studied by Kadem and co-workers(2015) [15]. Transmittance spectra of doped PEDOT:PSS layers increased to 6% and reached 94% in case of LiCl. Doping results into increase of surface roughness and electrical conductivity. The PCE of 6.82%, FF of 61% and Jsc of 18 mA.cm^{-2} were obtained.

Kim et al. (2010) [16] fabricated OSCs through all spray deposition of PEDOT:PSS and an active layer by substrate heating technique. The conductivity of PEDOT:PSS was enhanced about 10,000 times by addition of dimethylformamide (DMF) or DMSO. PEDOT:PSS transport layer having DMSO had PCE of 3%, Voc of 0.62 V and Jsc of 14.02 mA/cm^2. Transmittance of spin-coated and spray-deposited PH750, and spray-deposited PH750 containing 5% DMSO has been shown in Figure 2 [16]. The transmittance for spray-coated PEDOT:PSS between 300-500nm was 5% lower as compared to the spin-coated layer between 600-800 nm. Spray-deposited layer of PEDOT:PSS having 5% DMSO had transmittance of 90% and less than 85% in short and long wavelength range, respectively. The current density and voltage (J–V) curve show that the current densities of 6, 12 and 14 mA/cm^2 and PCE of 1.04%, 2.36% and 2.95% were obtained for pure PH750, PH750\5% DMF and PH750\5% DMSO layers, respectively (Figure 3) [16]. Very low fill factor for pure PH750 (0.267) reveals that it had high series resistance due to the presence of stacks of PEDOT:PSS droplets.

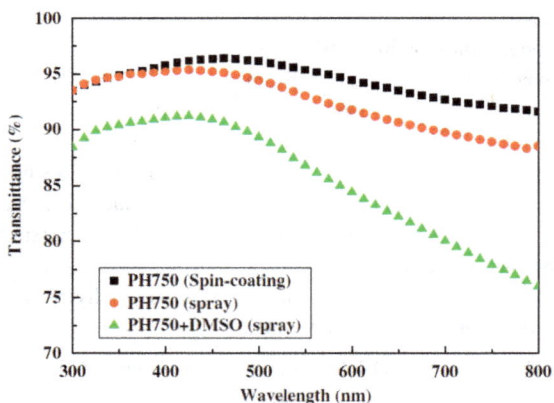

Figure 2 Transmittance of spin-coated PH750, spray-deposited PH750, and spray-deposited PH750\5% DMSO films (Reprinted with permission from Elsevier, [16]).

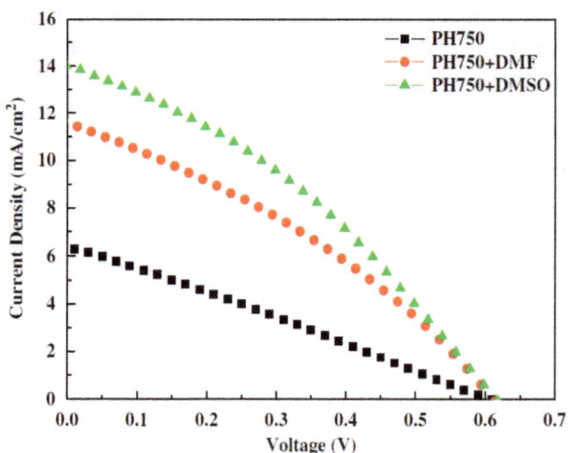

*Figure 3 J–V curve for PH750, PH750\5% DMF and PH750\5% DMSO layers
(Reprinted with permission from Elsevier, [16]).*

Oh and co-workers. (2013) [17] reported hybrid BHJ OSCs based on ZnO nanoparticles blended with P3HT and PCBM and having PEDOT:PSS or DMF modified PEDOT:PSS buffer layers. Reference device had a configuration of (ITO\PEDOT:PSS)\P3HT:PCBM\(LiF\Al) with an efficiency of 1.55% (Figure 4) [17]. The device (ITO\PEDOT:PSS(DMF) \ZnO:P3HT:PCBM\LiF\Al) having active layer with ZnO nanoparticle-doping and DMF modified PEDOT:PSS buffer layer delivered an efficiency of 3.39% owing to the synergistic effects. The low resistivity of PEDOT:PSS buffer layer and ZnO nanoparticles improved the Jsc and Voc value of OSCs, respectively. Energy band diagram of hybrid OSCs based on zinc (II) oxide (ZnO)-nanoparticles has been shown in Figure 5 [17]. DMF-modified PEDOT:PSS had work function similar to that of pristine PEDOT:PSS films. ZnO nanoparticles present in active layer facilitates the transport of electron carriers towards cathode due to high electron mobility as well as low conduction energy level. These electrons cannot be transferred to the anode due to the presence of hole transporting layer of PEDOT:PSS. Therefore, device performance was enhanced. DMF modulated PEDOT:PSS hole-transporting layer improved the electrical properties and had a resistivity of 1.45×10^{-2} $\Omega \cdot cm$ as compared to resistivity of 7.87×10^{-1} $\Omega \cdot cm$ for pristine PEDOT:PSS film. The J−V curves obtained under illuminations of A.M. 1.5 G with 100 mW/cm^2 intensity has been shown in Figure 6 [17]. The reference cell had an efficiency of 1.55%. ZnO nanoparticle doping improved the Voc and Jsc. AFM images show that addition of large sized ZnO nanoparticles to the active layer amplified the films roughness to 5.67 nm as compared to 1.22 nm for pristine P3HT:PCBM layer [Figure 7 (a) and (b)].

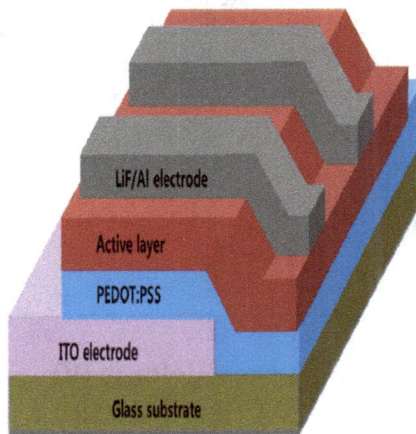

Figure 4 Structure of organic and ZnO nanoparticle-doped hybrid solar cell (Reprinted with permission from ACS publications, [17]).

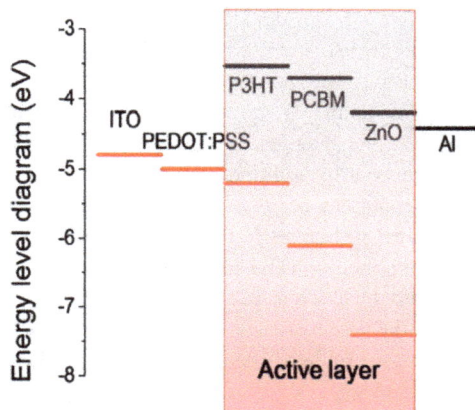

Figure 5 Schematic diagram of the band diagram for ZnO nanoparticles in the active layer (Reprinted with permission from ACS publications, [17]).

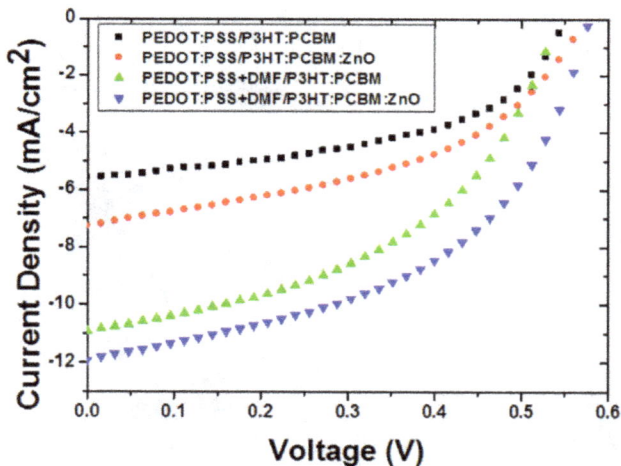

Figure 6 Current–voltage characteristics of P3HT:PCBM and ZnO:P3HT:PCBM thin films with PEDOT:PSS and DMF-modulated PEDOT:PSS (Reprinted with permission from ACS publications, [17]).

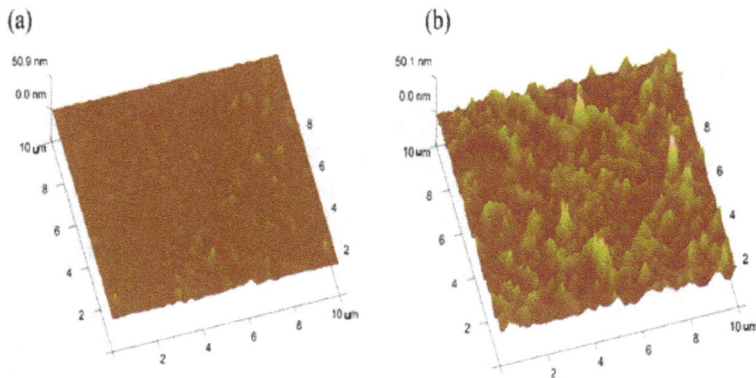

*Figure 7 AFM images of (a) P3HT:PCBM and (b) ZnO:P3HT:PCBM surfaces
(Reprinted with permission from ACS publications, [17]).*

2.3   PEDOT:PSS as buffer layer

Yoon et al. (2010) [18] added self-assembled Ag nanoparticle layer between PEDOT:PSS and P3HT:PCBM layer. This resulted in enhancement of short circuit current (Isc) along with decrease in Voc and FF because of the energy barrier due to the self assembled layer of Ag particles. Qiao et al. (2011) [19] used hydrophilic gold nanospheres having a diameter of 15 nm and 5 nm dispersed in the anode buffer layer of PEDOT:PSS. Localized surface plasmon resonance (LSPR) effect depending on the size of Au nanospheres improves the device efficiency. Enhancement of light harvesting by active layer in the region of Au nanospheres along with exciton generation and dissociation was observed. OSCs with 15 nm Au nanosphere doping had an efficiency of 2.36%. UV-ozone treated PEDOT:PSS was used as anode buffer layer in Cu phthalocyanine and fullerene based OSCs by Su et al. (2012) [20]. It had a PCE of about 20% due to improved work function of buffer layer. This leads to better contact between PEDOT:PSS and Cu phthalocyanine. Extraction efficiency of photo generated holes increases whereas electrons and holes recombination probability in active layer gets decreased. Long exposure to UV-ozone treatment declines the device performance due to saturation of work function change, breaking of chemical bonds, and ping hole defect formation in PEDOT:PSS layer leading to decrease in extraction efficiency of charge carriers.

Xi et al. (2010) [21] also studied OSCs based on Cu phthalocyanine and fullerene. PEDOT:PSS/LiF were used as double anode buffer layer which combats with stability

Materials for Solar Cell Technologies II   Materials Research Forum LLC
Materials Research Foundations **104** (2021) 40-76   https://doi.org/10.21741/9781644901410-3

issues of the device as compared to the devices based on single buffer layer. A Metal-Insulator-Semiconductor (MIS) model was put forward to explain the improvement in device performance as well as stability. PEDOT:PSS acts as a conductive organic ink, therefore, it behaves as a metal. CuPc and LiF are semiconductor and insulator, respectively. Therefore, PEDOT:PSS/LiF/CuPc system behaves as MIS model (Figure 8) [21]. Several holes pass from one side of 1 nm thick LiF layer to other by tunneling, when they come across the LiF/CuPc interface. High tunneling efficiency causes most of the holes to pass through the LiF layer. The few holes left get accumulated at LiF/CuPc interface. Due to capacitance like structure of MIS, electrons are induced to accumulate at PEDOT:PSS/LiF interface and build-in electric field acts towards PEDOT:PSS from CuPc. This field has direction parallel to that of the heterojunction, hence, Voc gets enhanced. LiF layer increases the interface quality and decreases the probability of recombination of excitons. Jsc can be further improved to a small extent due to pushing of holes across the LiF layer by the build-in field. Hole exporting efficiency is also enhanced due to smoothening of the ITO surface by PEDOT:PSS. The sheet resistance is improved due to hygroscopic nature of PEDOT:PSS [21-23].

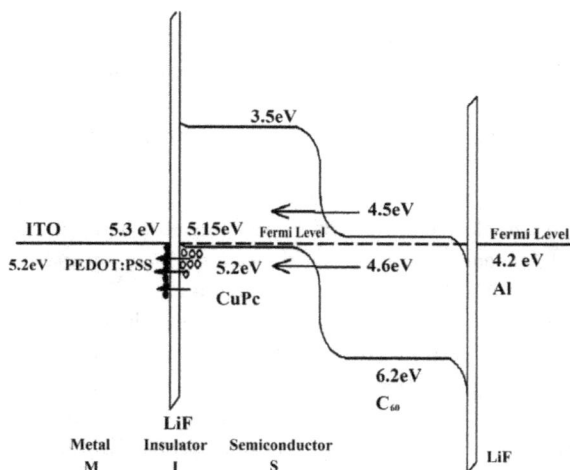

Figure 8 The MIS structure formed by PEDOT:PSS/LiF/CuPc system. Since the ITO electrode has been treated by UV-ozone, the work function can be enhanced a lot. It has been reported to increase to increase to over 5.2 eV. In this work, we consider the work function of ITO is enhanced to 5.2 eV after UV-ozone treatment (Reprinted with permission from Elsevier, [21]).

2.4 Inverted organic solar cells

Poor stability and organic semiconductor degradation by oxygen and moisture hampers the use of OSCs. Traditional OSCs lacking encapsulation are unstable due to etching of the ITO by acidic PEDOT:PSS along with easily oxidizable metal electrode having low work function like Al. As already stated, inverted OSCs (IOSCs) offer better stability because PEDOT:PSS is used above the active layer and it does not remain in contact with ITO. Better results can be obtained on using an air-stable metal electrode having high work function like Ag or Au. ITO and metal are generally used as the electron and hole collecting electrode. PEDOT:PSS is generally used as hole transport layer. Kuwabara et al. (2008) [24] prepared ITO/TiOx electrode through sol-gel method and used it in a device having ITO/TiOx/PCBM:P3HT/PEDOT:PSS/Au structure and an active area of 1 cm^2. PCE of 2.47% was attained under simulated sunlight irradiation (AM 1.5G-100mW.cm^2). BHJ IOSCs based on fluorine doped TiOx/electrodeposited amorphous (or anatase) TiOx electrode delivered PCE of 2.5% under AM 1.5-100 mW.cm^2 simulated sunlight irradiation [25]. Kang et al. (2012) [26] prepared IOSCs with a configuration of ITO/ZnO/P3HT:PCBM/PEDOT:PSS/Ag. The electron selective layer, active layer and hole selective layer were deposited through spray-coating. PCE of 3.17 and 1.33% were obtained for device area of 0.36 and 15.25 cm^2, respectively, when AM. 1.5 simulated illumination was used.

Various properties of solvents like wetting ability, surface tension, boiling point, etc. play important role in uniform coating of organic layers. Peh et al. (2011) [27] reported that AI4083 layer improves the wetting ability of P3HT:PCBM surface. Spray coating dynamics were studied through microscopic videos. Optimization of wetting ability and drying time for the suspension of PEDOT:PSS results into a transparent spray coated anode maintaining the device performance. Application of the vacuum-free method to near IR absorber yields device with 60% transparency. High resolution Ag grid embedded in PEDOT:PSS was used as a transparent hybrid electrode in large-area and high efficiency IOSCs. It has large transparency and small sheet resistance which can be minimized to about 1.2 Ω sq^{-1} through property tuning of Ag-grid. Device area of 1.21 cm^2 delivered a PCE of 3.36% and 5.85% for P3HT:PC61BM and PTB7:PC71BM devices, respectively.

Long term stability remains an issue in case of non-encapsulated OSCs. Stability can be improved by removing the ITO/PEDOT:PSS interface use of metal electrode having high work function [28]. In order to combat the wettability issue of hole transport layer of PEDOT:PSS, a fluoro surfactant, Capstone Dupont FS-31 was used as additive in place of Zonyl FS-300 by Lim and co-workers (2012) [29]. An efficiency of 3.1% was obtained along with stability for about 400 hours without encapsulation. Zimmerman and co-workers (2009) [30] studied the long term stability under sulphur plasma lamp $1000W/m^2$ illumination at 50 °C of inverted P3HT:PCBM solar cells. 2.5% PCE was obtained.

2.5 Tandem organic solar cells

Single unit OSCs lack practical device efficiency. Tandem OSCs have improved efficiency because the combination of two sub-cells using an interconnection layer allows utilization of wider solar spectrum range. The working of a tandem solar cell is as follows:

1. absorption of photons

2. generation of free carrier;

3. electron and hole collection from front and rear subcells, respectively by interconnecting layer.

4. electron and hole recombination in the interconnecting layer

5. charge collection of holes and electrons in the front and rear subcells, respectively by external circuit.

Step number 3 and 5 get reversed for an inverted tandem solar cell. An ideal serially connected tandem cell has Voc equal to additive sum of Voc of subcells and does not show potential loss in interconnecting layer. The subcells need to be optimizing order to gain best device performance. Liu et al. (2013) [31] PEDOT:PSS as the interconnecting layer in solution-processed homo-tandem device containing conjugated polyelectrolyte. Very high PCE of 10.1% was obtained due to the use of photoactive layer with double thickness. Lee et al. (2011) [32] used solution-processible nanoparticles of ZnO as n-type and MoO_3 or PEDOT:PSS as p-type material. P3HT:PCBM (higher band gap) and ZnPc/C60 (lower band gap) were used in the bottom and top cell, respectively. The ZnO/PEDOT:PSS inter-electrode, has Voc equal to the sum of upper and lower cell Voc, indicating the successful connection of the cells.

Li et al. (2017) [33] reported a solution-processed, tandem OSCs using small compounds ((5Z,5′E)-5,5′-((5″,5″″′-(4,8-bis((2-ethylhexyl)thio)benzo[1,2-b:4,5-b′]dithiophene-2,6-diyl)bis (3,3″ dioctyl-[2,2′:5′,2″-terthiophene]-5″,5-diyl))bis(methanylylidene))bis(3-ethyl-2-thioxo thiazolidin-4-one) (DR3TSBDT) and 5,15-bis(2,5-bis(2-ethylhexyl)-3,6-dithienyl-2-yl-2,5-dihydro-pyrrolo[3,4-c]pyrrole-1,4-dione-5′-yl-ethynyl)-10,20-bis(5-(2butyloctyl)thienyl) porphyrin zinc(II) (DPPEZnP-TBO). DR3TSBDT and DPPEZnP-TBO were used as electron donors in the upper and lower cells, respectively. A 40 nm thick hole transport layer of PEDOT:PSS was used. Extremely high PCE of 12.50% was obtained. Figure 9(a) [33] shows the molecular structure of DR3TSBDT and DPPEZnP-

TBO. Their corresponding absorption spectra [Figure 9(b)] shows that DR3TSBDT absorbs in the visible range extending upto 700 nm, whereas DPPEZnP-TBO absorbs the near-IR photons till 907 nm, thus covering the whole solar spectrum. The J-V curves of individual cells has been shown in Figure 9(c). DR3TSBDT:PC$_{71}$BM and TBO:PC$_{61}$BM based devices show Voc around 0.9 and 0.73 V, Jsc of about 14 and 18.53 mA.cm^{-2}, respectively. External quantum efficiency curves (EQE) DR3TSBDT and DPPEZnP-TBO based devices show high response in 300-650 and 300-900 nm range. The later showed appreciable EQE response in 650-900 nm range, whereas the former had weak EQE. This shows the balancing absorption nature of the donor materials used in the top and bottom cells.

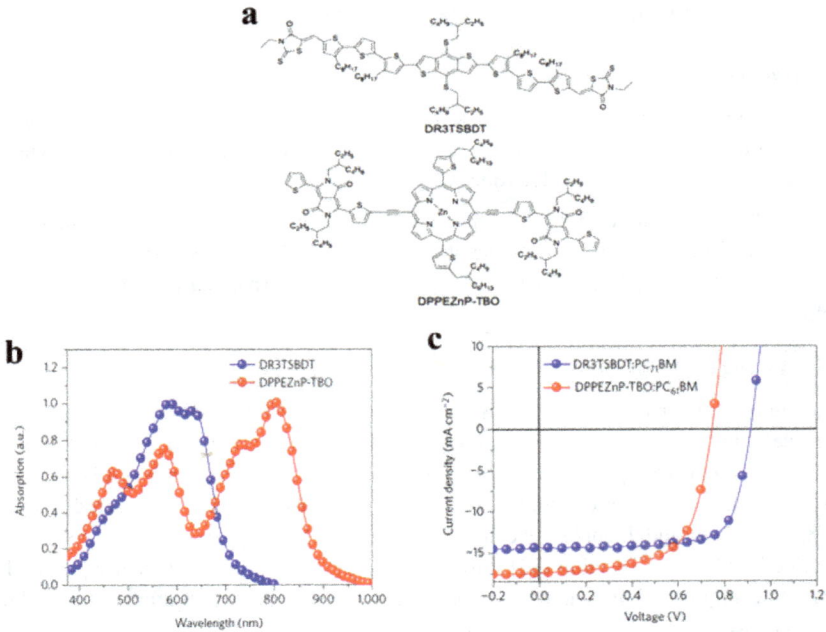

Figure 9 (a) Molecular structures of DR3TSBDT and DPPEZnP-TBO. (b) Normalized absorption spectra of DR3TSBDT and DPPEZnP-TBO films. (c) J–V curves (Reprinted with permission from Nature, [33]).

Materials for Solar Cell Technologies II   Materials Research Forum LLC
Materials Research Foundations **104** (2021) 40-76   https://doi.org/10.21741/9781644901410-3

Moet and co-workers (2010) [34] used middle electrode composed of PEDOT:PSS and ZnO in solution-processed tandem OSCs. PEDOT:PSS dispersion was pH modified which results into lowering of PEDOT:PSS work function. The polythiophene: fullerene based device performance remains unaffected but low Voc is obtained for polyfluorene derivative based device due to its larger ionization potential. When perfluorinated ionomer layer is used, anode work function gets recovered and a Voc of 1.92 V is obtained for polyfluorene-based double junction OSCs. Zhou et al. (2012) [35] prepared an inverted tandem polymer solar cells containing PEDOT:PSS. They modified one of the interface using high performance ethoxylated polyethylenimine (PEIE) charge recombination layer having low absorption, high conductivity and front and rear interface work function contrast (1.3 eV). It produces an ideal tandem cell having Voc equal to sum of subcells. But, the fill factor is higher than that of the individual subcells.

3   Dye sensitized solar cells

DSSCs mainly consist of: 1) porous oxide film, 2) dye sensitized photo anode (mesoporous TiO_2 absorbed with molecule of dye), 3) liquid electrolyte of triiodide/iodide, 4) Pt counter electrode (CE).

PEDOT:PSS in its bare or hybrid form is mainly used to enhance the catalytic activity of CE. In this section, we will discuss the effect of using PEDOT:PSS with graphene, CNT, TiS_2, TiO_2 (nanoparticle), SiO_2 (nanoparticle), CoS (nanoparticles) etc. on the performances of CE.

3.1   PEDOT:PSS as counter electrode

The role of counter electrode (CE) is to collect electron and catalyse redox reaction of triiodide/iodide. Most of the DSSCs consist of Pt on fluorine doped tin oxide glass. Pt exhibits efficient catalytic activity for triiodide/iodide redox reaction but it is expensive too. To develop cheaper CE replacement of this expensive metal is required [36]. In 1998, Yohannes and Inganäs revealed that the PEDOT may serve as the catalyst for triiodide/iodide redox reaction [37]. PEDOT:PSS CE were prepared recently [38] and its excellent catalytic activity and its low cost attracted researchers from around the world.

Ma et al. (2018) [39] prepared composite of reduced graphene oxide (rGO) with PEDOT and applied it as CE for DSSCs. Photoelectric conversion efficiencies of 7.79, 8.33 and 6.88% were observed for rGO/PEDOT, Pt and PEDOT, respectively under the same experimental conditions. Xia et al. (2007) [40] studied electrochemical and catalytic activity of PEDOT CE with different counter ions like ClO_4^-, TsO^- and PSS. PEDOT-ClO_4^- system exhibits better catalytic performances due to higher porosity and smaller

Materials for Solar Cell Technologies II   Materials Research Forum LLC
Materials Research Foundations **104** (2021) 40-76   https://doi.org/10.21741/9781644901410-3

size of ClO_4^- ion..In DSSCs PEDOT:PSS film is a good option for cathodic catalyst. Kitamura and Shiratori (2011) [41] prepared thin film of PEDOT:PSS using layer-by-layer self-assembly method (LbL) for the first time. The as prepared film exhibits highly adhesive nature and its thickness could be manipulated up to the controlled scale. Authors used carbon black as an additive in PEDOT:PSS solution and fabricated extremely mesoporous film. As expected, this film exhibits outstanding catalytic properties as cathode in DSSCs. Authors revealed morphological changes by varying carbon black content and studied the effect of porosity on its performances as a CE. They observed that the film exhibits 8% lower efficiency as compared to the conventional counter electrode.

Carbon nanotube (CNT) fiber is well known for its outstanding mechanical and electronic properties at nanoscalar as well as microscopic scale. The fiber displays conductivity of 103 Scm^{-1} and its tensile strength is of the order of 10^3 MPa.

Guozhen Guan et al. (2013) [42] fabricated CNT–PEDOT:PSS composite from 1:1 concentration of PEDOT:PSS and CNT. In their experiment they observed that the sheet thickness of 2 nm composite exhibits enhanced FF with increasing thickness of CNT and decreasing polymer percentage. However V_{OC} and J_{SC} did not vary much. Aligned CNT sheets help to induce better catalytic activity of bare PEDOT:PSS through p-p interaction. The authors also revealed that the aligned composite of CNT-PEDOT:PSS exhibits the maximum conversion efficiency of 8.3% when it were used as electrodes for DSSCs. Lee et al. (2015) [43] used graphene dots in PEDOT: PSS and observed rough surface morphology, better conductivity and low charge transfer resistance for this CE. Better cell efficiency was also observed and was

found to be 43% higher than the cell containing PEDOT: PSS. PEDOT: PSS performed well with graphene at low light which makes it suitable for indoor use.

Li et al. (2013) [44] compared morphology of different CEs and found high efficiency (7.04%) for TiS_2/PEDOT:PSS composite due to the larger surface area of PEDOT:PSS. They revealed that TiS_2 particle acts as an electrocatalyst material in redox reaction. Furthermore, the composite serves as catalyst for the conversion of I^- to I_3^-. They demonstrated that this cell could replace the expensive Pt counter electrode based cells. Maiaugree et al. (2012) [45] also incorporated TiO_2 nanoparticles in PEDOT:PSS and observed better catalytic activity as well as energy conversion efficiency for the composite and reduced charge transfer resistance for CE. Its improved catalytic activity has been proved through cyclic voltammogram. Reduction peaks increased by 0.11 mA after incorporation of TiO_2 nanoparticles in PEDOT:PSS. Muto and co-workers (2007) [46] incorporated nano crystals of TiO_2 in PEDPT:PSS. They observed better cathodic activity and energy conversion efficiency of about 4.38%. Cyclic voltammogram studies revealed that the TiO_2 added PEDOT:PSS has highest reaction rate than PEDOT:PSS. 4 wt% of TiO_2 exhibits maximum cathodic activity due to improved surface area and catalytic/conductive utility of PEDOT:PSS.

Xu et al. (2012) [47] observed better electrocatalytic activity and electrical conductivity when TiN was incorporated in PEDOT:PSS. This composite provides better and efficient interfacial active sites resulting into energy conversion efficiency of 7.06% which is higher than that of the cell having Pt-FTO (6.57%) as CE. Cyclic voltammetric (Figure 10 [47]) analysis shows that the TiN nanorods and nanoparticles based composites with PEDOT:PSS exhibits electron transfer rate comparable to Pt-FTO. Tin mesoporous sphere based composites with PEDOT:PSS exhibits lower electron transfer rate than the Pt-FTO. TiN based composites with PEDOT exhibits better current density (Figure 11 [47]) than the TiN based electrodes. Yeh et al. (2011) [48] observed higher energy conversion efficiency (6.67%) than the counter electrode having s-Pt (6.57%). The counter electrode of TiN nanoparticles with PEDOT:PSS exhibits better catalytic activity, lower charge transfer resistance between CE and electrolyte [49]. TiN-NPs/PEDOT:PSS film provides multifold pathway for electron transfer due to continual connection of TiN-NPs with PEDOT:PSS (Figure 12 [49]). Transmittance at longer wavelengths increased when SiO_2 was added to PEDOT:PSS due to decreased anisotropy and interaction of PEDOT-PSS, as a result of homogeneous mixing of SiO_2 nanoparticles with PEDOT-PSS. Features like mesoporous (Figure 13 [49]) and larger electrochemical surface area results into better catalytic activity of SiO_2/PEDOT-PSS CE (Figure 14 [49]).

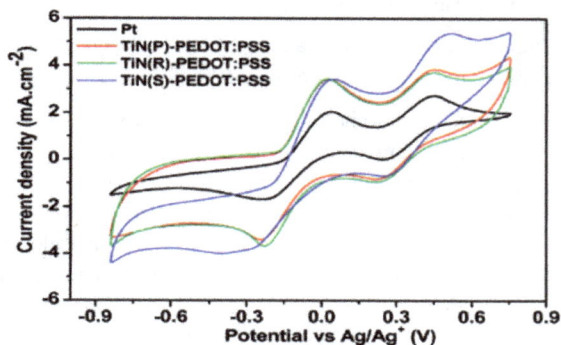

Figure 10 Cyclic voltammograms of Pt, TiN(P)-PEDOT:PSS, TiN(R)-PEDOT:PSS, TiN(S)-PEDOT:PSS counter electrodes in 10 mM LiI, 1 mM I2 and 0.1 M LiClO4 acetonitrile solution at a scan rate of 20 mV s-1 (Reprinted with permission from ACS publication, [47]).

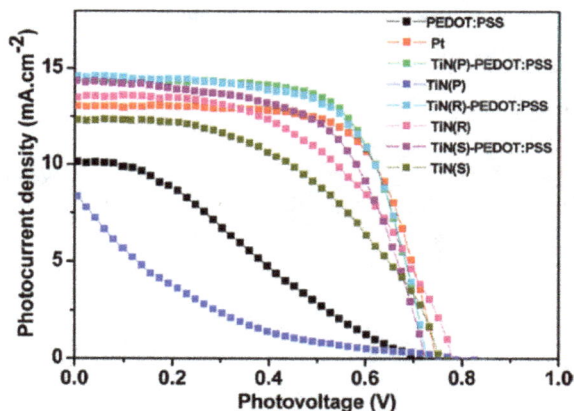

Figure 11 Characteristic photocurrent density-voltage (J-V) curves of DSSCs with different electrodes, measured under simulated sunlight 100 mW cm-2 (AM 1.5). The liquid electrolyte is composed of 0.05 M I2, 0.1 M LiI, 0.6 M 1,2-dimethyl-3-propylimidazolium iodide (DMPII), and 0.5 M 4-tert-butyl pyridine in acetonitrile solution (Reprinted with permission from ACS publication, [47]).

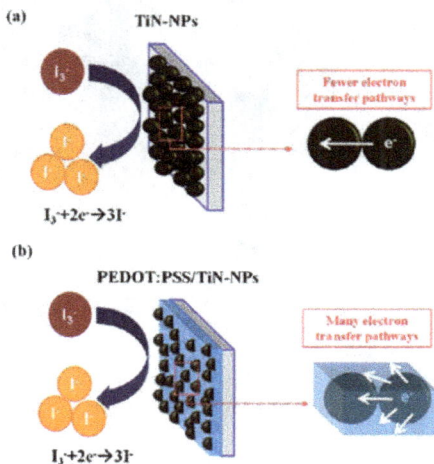

Figure 12   (a) Electron transfer pathways in a film of TiN-NPs and (b) electron transfer pathways in a composite film of PEDOT:PSS/TiN-NPs (Reprinted with permission from Royal Society of Chemistry, [48]).

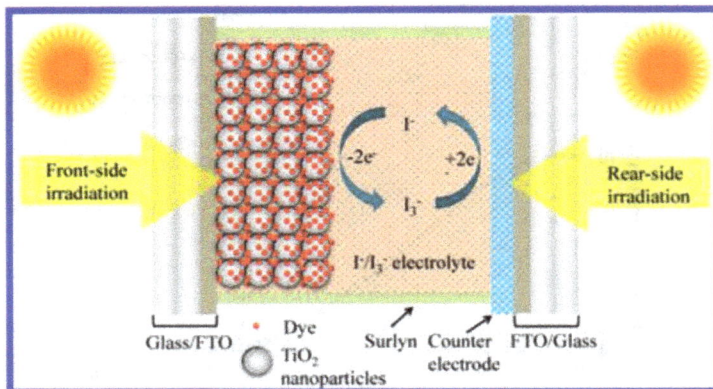

Figure 13 Schematic device structure of bifacial DSCs (Reprinted with permission from ACS publication, [49]).

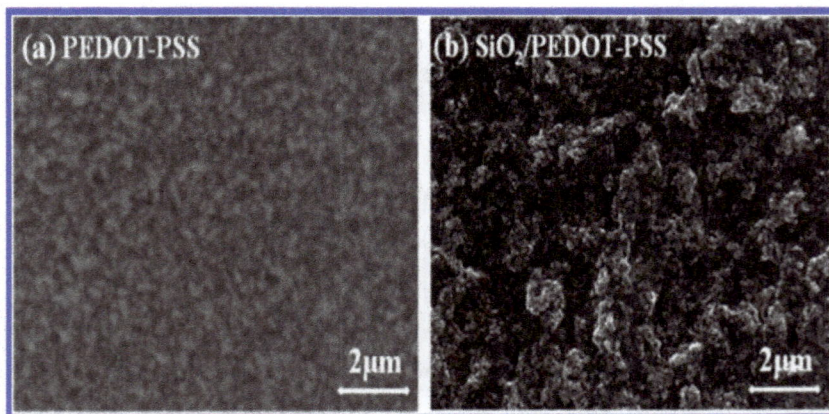

Figure 14 SEM images from (a) PEDOT-PSS film and (b) SiO2 (40 mg)/PEDOT-PSS film (Reprinted with permission from ACS publication, [49]).

Sudhagar et al. (2011) [50] incorporated nanoparticle of cobalt sulphide into PEDOT:PSS and the synergistic effect resulted into better catalytic activity. The reduction of triiodide was also observed to be easier than the pristine PEDOT:PSS system. CoS nanoparticles form two-dimensional sheet like structure and PEDOT:PSS serves as filler agent which results into better contact between triiodide ions and PEDOT:PSS. Cyclic voltammogram (Figure 15 [50]) was also studied for various electrodes.

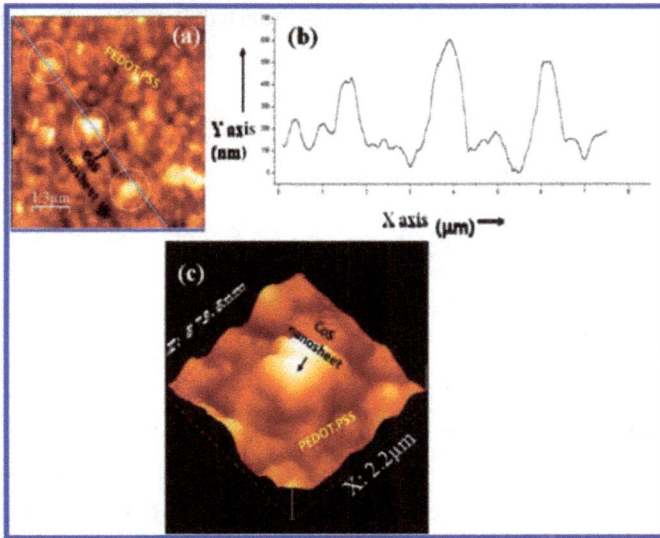

Figure 15 (a) AFM images of the sheetlike CoS dispersed in PEDOT:PSS and (b) surface profile of the sheetlike CoS on PEDOT:PSS matrix, and (c) CoS/PEDOT:PSS interface (Reprinted with permission from ACS publication, [50]).

Zhang et al. (2012) [36] incorporated nanocrystals of $CuInS_2$ with PEDOT:PSS and applied it as CE for DSSCs. Authors observed better catalytic activity for the composite as compared to PEDOT:PSS or $CuInS_2$. Photoelectric conversion efficiencies were reported to be 6.50, 6.51, 5.45, and 3.22 % for the composite, Pt, $CuInS_2$ and PEDOT:PSS, respectively. Maiaugree et al. (2018) [51] prepared composite of $SrTi_{1-x}Co_xO_3$ (x = 0, 0.025, 0.05, 0.075 and 0.1) nanoparticles with PEDOT:PSS and applied it as CE for DSSCs. Outstanding catalytic performances were observed for iodide/triiodide redox reaction. Furthermore, authors observed highest solar conversion efficiency of 8.39% for the composite having composition of $SrTi_{0.925}Co_{0.075}O_3$ NPs/PEDOT-PSS as compared to 8.27% for device having Pt as CE. In Figure 16 [51], indicated plane in brackets are the plane for pervoskite structure. XRD of $SrTiO_3$ nanoparticles reveals that the impurities are absent and Ti sites are perfectly substituted by Co. The lowest R_{ct} value for $x = 0.075$ is observed which is associated to the maximum catalytic activity of $SrTi_{0.925}Co_{0.075}O_3$ NPs/PEDOT-PSS film (Figure 17, Figure 18 [51]). It can be concluded that the conductivity and electrocatalytic activity can be affected by doping of Co in $SrTiO_3$ nanoparticles. The enhancement of electrocatalytic activity and conductivity can be achieved by varying Co content.

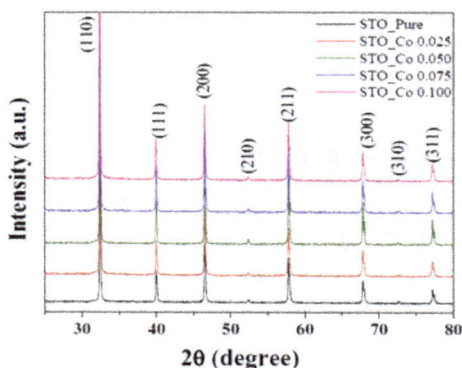

Figure 16 XRD patterns of $SrTi_{1-x}Co_xO_3(x=0, 0.025, 0.05, 0.075$ and 0.1) NPs (Reprinted with permission from Elsevier,[51]).

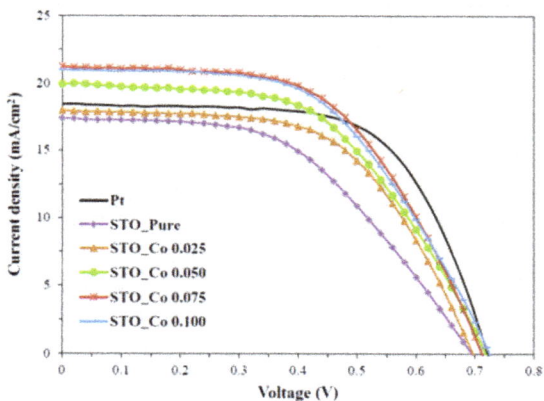

Figure 17 Photocurrent density (J) vs, photovoltage (V) curves of DSSCs with based on STCoO NPs/PEDOT films (Reprinted with permission from Elsevier,[51]).

Materials Research Forum LLC
https://doi.org/10.21741/9781644901410-3

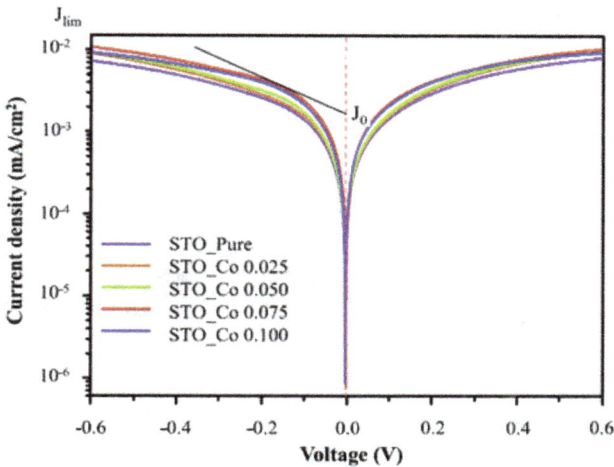

Figure 18 Tafel plots of STCoO NPs/PEDOT-PSS CEs (Reprinted with permission from Elsevier,[51]).

Xu et al. (2015) [52] applied PEDOT:PSS film and ten modified films of PEDOT:PSS as CE for the study of biafacial DSSCs. PEDOT:PSS film showed front and rear side efficiencies of 4.71 and 2.38% respectively. They used different additives like iodine, DMSO and PEG (molecular wt. 200) and studied optical properties (Figure 19 [52]) of modified PEDOT:PSS film and J-V curve (Figure 20 [52]) and demonstrated enhanced efficiencies of 5.19, 5.11 and 5.29 % respectively for front side due to poorer sheet resistance. Furthermore, the efficiencies were obtained almost unchanged for rear side illumination. Ahmed et al. (2018) [53] fabricated Si_3N_4/MoS_2-PEDOT:PSS and applied as CE for biafacial DSSCs. They observed better catalytic activity for the CE than that of the pristine Si_3N_4/MoS_2 and PEDOT:PSS. The conversion efficiencies (PCE) for Si_3N_4/MoS_2-PEDOT:PSS (5% composition in CE), Si_3N_4/MoS_2 or PEDOT:PSS were observed to be 7.16, 3.80 and 4.20%, respectively. Song (2014) [49] added SiO_2 in PEDOT:PSS and used it as CE in bifacial DSSC. Better power conversion efficiency (4.61%) was observed for the rear side irradiation along with enhanced catalytic performance as compared to the bare PEDOT:PSS.

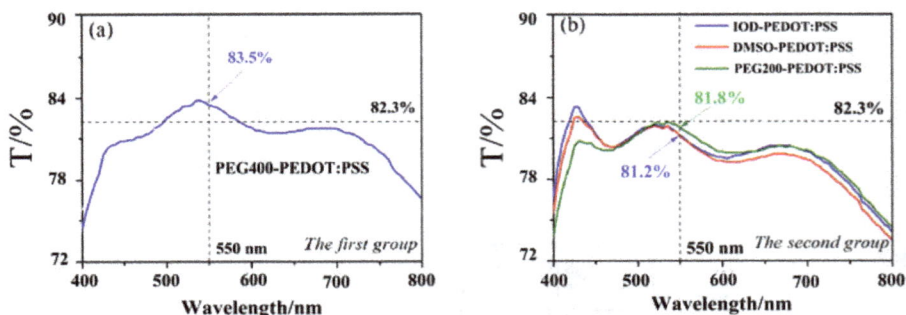

Figure 19 UV–Vis spectra of the different modified PEDOT:PSS films on the FTO substrates in thevisible range. (a) the first group including a modified film (PEG400-PEDOT:PSS), showingslightly enhanced transparency; (b) the second group including six modified films,showing almost unchanged transparency (Reprinted with permission from Elsevier,[52]).

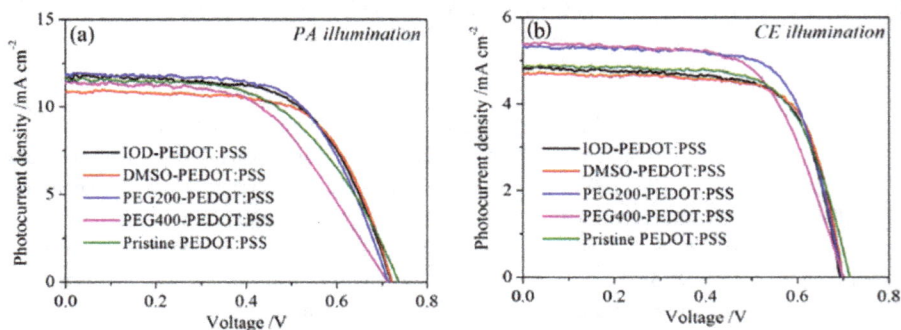

Figure 20 J-V curves of bifacial DSSCs using different PEDOT:PSS films as counter electrodes.(a) under photoanode (PA) illumination; (b) under counter electrode (CE) illumination (Reprinted with permission from Elsevier,[52]).

4. Hybrid silicon solar cell

Today, the solar cell market is primarily dominated by silicon wafer solar cells. The main challenges include reduction of production cost, maintaining the efficiency, avoiding high temperature fabrication, avoiding vacuum processes.

Materials for Solar Cell Technologies II   Materials Research Forum LLC
Materials Research Foundations **104** (2021) 40-76   https://doi.org/10.21741/9781644901410-3

To overcome all the above mentioned issues PEDOT:PSS is being used as hetero-emitter. Cost friendliness may be introduced through replacing crystalline Si with wafer Si. PEDOT:PSS forms heteroconjunction with n type Si. However, flexibility of Si hybrid solar cell is a major issue to be resolved. PEDOT with high work function (5.0 eV) serves as hole transporter. In this section we will go through some details regarding how PEDOT:PSS manipulates the performance of these hybrid solar cells.

Pietsch et al. (2014) [54] studied the variation of substrate doping concentration (N_d) and open circuit voltage (Voc). 645 mV of Voc was observed for Nd= 2.6×10^{17} and 12.6% of PCE for Nd = 2.5×10^{16}. Metallic conductivity of PEDOT:PSS were probably responsible for the high PCE and efficient transport of holes for planar hybrid n-Si/PEDOT:PSS solar cells. Thomas et al. (2016) [55] added co-solvent of methanol and ethylene glycol in PEDOT:PSS for better conductivity. When 16 wt% of the co-solvent was added, the so prepared solar cell delivered maximum PCE of 14.6% and open circuit voltage of 620 mV. This is probably due to the ordered grain structure at the macromolecular level. Benzenoid chains were ordered and quinoid chains were found disordered in which segregated PSS chains were present. Microstructural manipulation of PEDOT:PSS attributes better performances for Si hybrid solar cell .

He et al. (2011) [56] fabricated highly efficient hybrid solar cell through silicon nanowires (SiNWs) and PEDOT:PSS. Longer SiNW leads to reduced surface coverage of PEDOT and the PCE for Si/PEDOT:PSS and SiNW/PEDOT:PSS cell were found to be 6.2 and 9%, respectively.

Si nanoholes (SiNH) were studied by Hong et al. (2013) [57]. Planar Si shows high reflectance due to the huge variation of refractive indices between Si and air. On the contrary, SiNH exhibits lower reflectance. Moiz et al. (2012) [58] prepared PEDOT:PSS–silicon nanowire based hybrid solar cell. Sound linking and linear alignment of PEDOT:PSS on n-Si are the reason for enhanced efficiency, better short-circuit current and open-circuit voltage of the cell. Polymer thickness more than 70 nm leads to increased resistance as well as decreased electric field between the electrodes. Authors successfully prepared stamped channel of PEDOT:PSS of lesser thickness and overcame these two factors. Low mobility of free carriers in bulk polymer also increases the carrier recombination and lower the cell efficiency. Whereas stamped polymer provides lower surface area to interact with SiNW which consequently decreases the recombination losses and current leakage. Thin (about 10 nm) stamped channel of PEDOT:PSS also improves the extinction decay losses hence improves PCE. Park et al. (2015) [59] also prepared SiNW/PEDOT:PSS hybrid solar cell and observed 13.2% PCE. Using SiNW/PEDOT:PSS interface for hybrid solar cell leads to better stability. When ITO is used with PEDOT:PSS, degradation is a major issue. Authors used Au (200-μm pitch) mesh for the front electrode and observed excellent PCE (13.2%), FF (67.8%), short-circuit current density, Jsc (36.03 mA/cm^2) and open-circuit voltage, Voc (539.2 mV) as compared to the ITO. Syu and co-workers (2012) [60] prepared SiNW/PEDOT:PSS hybrid solar cell. They observed better light trapping and poor light reflectance for the cell as compared to the planar Si/PEDOT:PSS solar cell. PCE for less than 0.5 micrometer SiNWs was observed to be 8.40% for the cell. Sharma et al. (2014) prepared hybrid solar cell of SiNW arrangement and PEDOT:PSS and obtained a PCE of 6.62% [61]. Junghanns et al. (2015) [62] fabricated hybrid solar cells using multicrystalline Si and PEDOT:PSS. They studied the effect of SiO$_x$ and aluminium (III) oxide (Al$_2$O$_3$) on surface of Si. PCE were reported to be 10.3 and 7.3% for mc-Si/Al$_2$O$_3$/PEDOT:PSS and mc-Si/SiO$_x$/PEDOT:PSS, respectively (Figure 21 [62]).

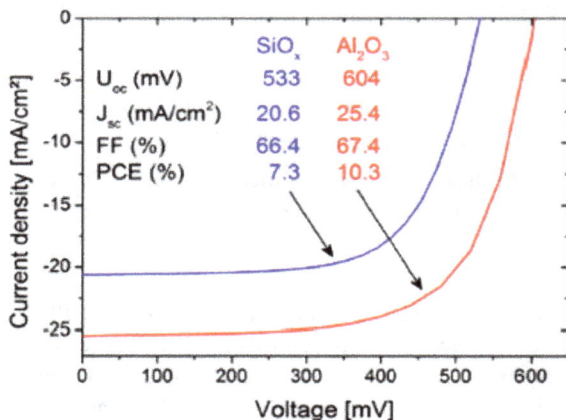

Figure 21   I-V characteristics of the hybrid solar cells with 1 nm SiOx (blue) and 0.75nm Al2O3 (red) as a passivation layer (Reprinted with permission from AIP publishing, [62]).

Si/PEDOT:PSS core/shell nanowire arrays were prepared by Lu et al. (2011) [63]. PCE was obtained to be 6.35% for the Si/PEDOT:PSS core/shell nanowire solar cell (Figure 22, Figure 23 [63]). Nagamatsu et al. (2014) [64] prepared silicon/PEDOT:PSS heterojunction through spin coating technique below 100 °C for blocking electron. PCE was observed to be 11.7% at AM1.5 illumination. PEDOT with PSS forms macromolecular salt in which oxidation takes place on PEDOT and reduction takes place on PSS forming cation and anion, respectively. PEDOT:PSS gets easily dispersed in aqueous medium, which can be further spin casted. Furthermore, as-obtained film of PEOT:PSS is highly transparent as well as conductive. PEDOT thin film serves as the doped p-type semiconductor.

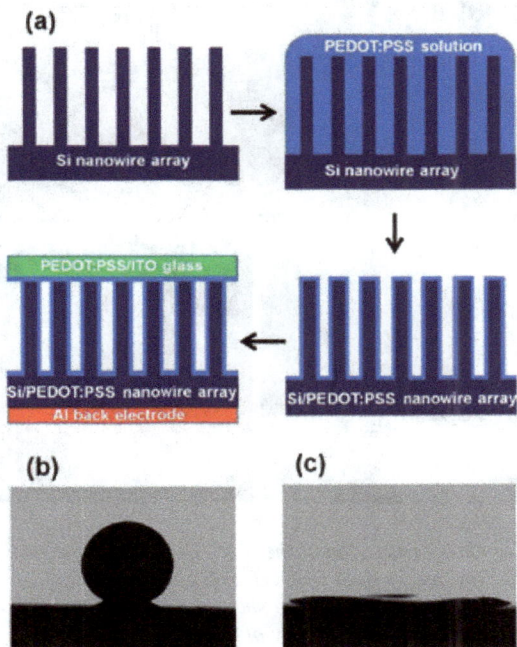

Figure 22 (a) Schematic illustration of the fabrication process of Si/ PEDOT:PSS core/shell nanowire array solar cells; photographs of (b) aspurchased and (c) ethanol diluted PEDOT:PSS solution on Si nanowire arrays, indicating different wettability (Reprinted with permission from Royal Society of Chemistry,[63]).

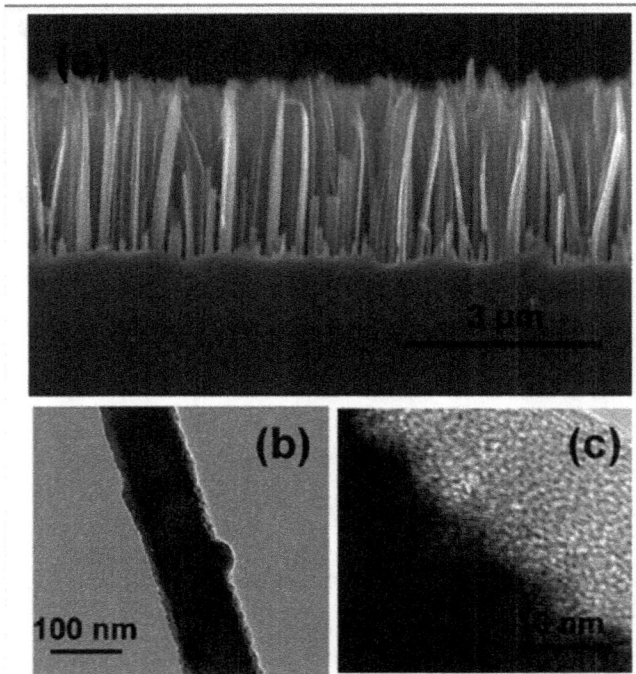

Figure 23 Structural characterization of the Si/PEDOT:PSS core/shell nanowire arrays. (a) Cross-sectional SEM image of Si/PEDOT:PSS core/ shell nanowire arrays; (b) and (c) TEM images of a Si/PEDOT:PSS core shell nanowire (Reprinted with permission from Royal Society of Chemistry,[63]).

Sheng et al. (2014) [65] improved wettability of Si surface through oxidizing agents nitric acid (HNO₃) and hydrogen peroxide (H₂O₂).. Treatment with these oxidizing agents enhanced surface coating of Si with PEDOT:PSS. The better heterojunction of inorganic-organic hybrid solar cell was prepared. Dry etching technique on Si was used to prepare Si nanocones/PEDOT SC. Monolayer polystyrene (PS) nanospheres were masked on Si/PEDOT SC. Outstanding light trapping and 7.1% of PCE were observed [66]. Zielke et al. (2014) [67] prepared 'back PEDOT' cell by using PEDOT:PSS at the rear side of the cell. Li et al. (2015) [68] prepared flexible Si/PEDOT:PSS hybrid solar cells. They demonstrated schematic preparation of solar cell with micro pyramidal Si and reported the PCE for planar flexible cell and micro pyramidal Si/PEDOT:PSS solar cell to be 4 and 6.3%, respectively.

Conclusion

PEDOT:PSS has been widely used in OSCs, inverted OSCs, DSSCs for improving performances of these solar cells. Vital application of combination of PEDOT:PSS with different materials is continuously attracting researchers from all over the world. Fullerene and fullerene based derivatives, single walled carbon nanotube, nanomaterials, metallic nanowires, etc. have been used to improve the performance of PEDOT:PSS in solar cells. The effect of additives like ethylene glycol, DMSO and various post-treatment has also been observed. It has been observed that inverted OSCs have better stability as compared to OSCs. Tandem OSCs use a combination of two sub-cells offer better efficiency and wider solar spectrum can be utilized. Combination of PEDOT:PSS with various counter ions, nanocrystals of $CuInS_2$, composite of $SrTi_{1-x}Co_xO_3$, nanoparticle of CoS, nanoparticle of SiO_2, nanorods of TiN, nanocrystals of TiO_2, nanoparticles of TiO_2, CNT fibre etc. improved the performance of CE for DSSCs. Similarly, combination of PEDOT:PSS with SiNW, SiNH etc. enhances the performance of Si based hybrid solar cells. There is a vast scope to imply PEDOT:PSS in above discussed solar cells.

References

[1] M. Kemerink S. Timpanaro, M. M. de Kok, E. A. Meulenkamp, F. J. Touwslager Three-dimensional inhomogeneities in PEDOT:PSS films Phys. Chem. B 2004, 108, 49, 18820-18825. https://doi.org/10.1021/jp0464674

[2] K. Kawano, N. Ito, T. Nishimori, J. Sakai, Open circuit voltage of stacked bulk heterojunction organic solar cells, Appl. Phys. Lett. 88 (2006) 073514. https://doi.org/10.1063/1.2177633

[3] H. Hoppe, N.S. Sariciftci, D. Meissner, Optical constants of conjugated polymer/fullerene based bulk-heterojunction organic solar cells, Mol. Cryst. Liq. Cryst. 385 (2002) 113-119. https://doi.org/10.1080/713738799

[4] M.G. Kang, M.S. Kim, J. Kim, L.J. Guo, Organic solar cells using nanoimprinted transparent metal electrodes, Adv. Mater. 20 (2008) 4408–4413. https://doi.org/10.1002/adma.200800750

[5] M.G. Kang, H.J. Park, S.H. Ahn, L.J. Guo, Transparent Cu nanowire mesh electrode on flexible substrates fabricated by transfer printing and its application in organic

solar cells, Sol. Energy Mater. Sol. Cells. 94 (2010) 1179–1184.
https://doi.org/10.1016/j.solmat.2010.02.039

[6] M. Kaltenbrunner, M.S. White, E.D. Głowacki, T. Sekitani, T. Someya, N.S. Sariciftci, S. Bauer, Ultrathin and lightweight organic solar cells with high flexibility, Nat. Commun. 3 (2012) 770. https://doi.org/10.1038/ncomms1772

[7] D. Qu, F. Liu, Y. Huang, W. Xie, Q. Xu, Mechanism of optical absorption enhancement in thin film organic solar cells with plasmonic metal nanoparticles, Opt. Express. 19 (2011) 24795-24803. https://doi.org/10.1364/OE.19.024795

[8] E. Kymakis, G. Klapsis, E. Koudoumas, E. Stratakis, N. Kornilios, N. Vidakis, Y. Franghiadakis, Carbon nanotube/PEDOT:PSS electrodes for organic photovoltaics, Eur. Phys. J. Appl. Phys. 36 (2007) 257–259. https://doi.org/10.1051/epjap:2006148

[9] Z. Li, F. Qin, T. Liu, R. Ge, W. Meng, J. Tong, S. Xiong, Y. Zhou, Optical properties and conductivity of PEDOT:PSS films treated by 4 polyethylenimine solution for organic solar cells, Org. Electron. 21 (2015) 144-148. https://doi.org/10.1016/j.orgel.2015.03.010

[10] H. Park, J.A. Rowehl, K.K. Kim, V. Bulovic, J. Kong, Doped graphene electrodes for organic solar cells, Nanotechnology. 21 (2010) 505204. https://doi.org/10.1088/0957-4484/21/50/505204

[11] Y.J. Noh, S.S. Kim, T.W. Kim, S.I. Na, Cost-effective ITO-free organic solar cells with silver nanowire–PEDOT:PSS composite electrodes via a one-step spray deposition method, Sol. Energy Mater. Sol. Cells. 120 (2014) 226–230. https://doi.org/10.1016/j.solmat.2013.09.007

[12] S. Park, S.J. Tark, D. Kim, Effect of sorbitol doping in PEDOT:PSS on the electrical performance of organic photovoltaic devices, Current Applied Physics 11 (2011) 1299-1301. https://doi.org/10.1016/j.cap.2011.03.061

[13] A. Singh, M. Katiyarab, A. Garg, Understanding the formation of PEDOT:PSS films by ink-jet printing for organic solar cell applications, RSC Adv. 5 (2015) 78677-78685. https://doi.org/10.1039/C5RA11032G

[14] T. Hori, Y. Miyake, N. Yamasaki, H. Yoshida, A. Fujii, Y. Shimizu, M. Ozaki, Solution processable organic solar cell based on bulk heterojunction utilizing phthalocyanine derivative, Appl. Phys. Express. 3 (2010) 101602. https://doi.org/10.1143/APEX.3.101602

[15] B. Kadem, W. Cranton, A. Hassan, Metal salt modified PEDOT:PSS as anode buffer layer and its effect on power conversion efficiency of organic solar cells, Org. Electron. 24 (2015) 73–79. https://doi.org/10.1016/j.orgel.2015.05.019

[16] K.J. Kim, Y.S. Kim, W.S. Kang, B.H. Kang, S.H. Yeom, D.E. Kim, J.H. Kim, S.W. Kang, Inspection of substrate-heated modified PEDOT:PSS morphology for all spray deposited organic photovoltaic's, Sol. Energy Mater. Sol. Cells. 94 (2010) 1303–1306. https://doi.org/10.1016/j.solmat.2010.03.013

[17] S.H. Oh, S.J. Heo, J.S. Yang, H.J. Kim, Effects of ZnO Nanoparticles on P3HT:PCBM Organic Solar Cells with DMF-Modulated PEDOT:PSS Buffer Layers, ACS Appl. Mater. Interfaces. 5 (2013) 11530-11534. https://doi.org/10.1021/am4046475

[18] W.J. Yoon, K.Y. Jung, J. Liu, T. Duraisamy, R. Revur, F. L. Teixeira, S. Sengupta, P.R. Berger, Plasmon-enhanced optical absorption and photocurrent in organic bulk heterojunction photovoltaic devices using selfassembled layer of silver nanoparticles. Sol. Energy Mater. Sol. Cells. 94 (2010) 128–132. https://doi.org/10.1016/j.solmat.2009.08.006

[19] L. Qiao, D. Wang, L. Zuo, Y. Ye, J. Qian, H.Z. Chen, S. He, Localized surface plasmon resonance enhanced organic solar cell with gold nanospheres, Appl. Energy. 88 (2011) 848–852. https://doi.org/10.1016/j.apenergy.2010.09.021

[20] Z. Su, L. Wang, Y. Li, H. Zhao, B. Chu, W. Li, Ultraviolet-ozone-treated PEDOT:PSS as anode buffer layer for organic solar cells, Nanoscale Res. Lett. 7 (2012) 465. https://doi.org/10.1186/1556-276X-7-465

[21] X. Xi, Q. Meng, F. Li, Y. Ding, J. Ji, Z. Shi, G. Li, The characteristics of the small molecule organic solar cells with PEDOT:PSS/ LiF double anode buffer layer system, Solar Energy Materials & Solar Cells 94 (2010) 623–628. https://doi.org/10.1016/j.solmat.2009.12.014

[22] R. Pacios, A.J. Chatten, K. Kawano, J.R. Durrant, D.D.C. Bradley, J. Nelson, Effects of photo-oxidation on the performance of poly[2-methoxy-5-(3,7-dimethyloctyloxy)-1,4-phenylene vinylene]:[6,6]-phenyl C61-butyric acid methyl ester solar cells, Adv. Funct. Mater. 16 (2006) 2117–2126. https://doi.org/10.1002/adfm.200500714

[23]   K. Kawano, R. Pacios, D. Poplavskyy, J. Nelson, D.D.C. Bradley, J.R. Durrant, Degradation of organic solar cells due to air exposure, Sol. Energy Mater. Sol. Cells. 90 (2006) 3520–3530. https://doi.org/10.1016/j.solmat.2006.06.041

[24]   T. Kuwabara, T. Nakayama, K. Uozumi, T. Yamaguchi, K. Takahashi, Highly durable inverted-type organic solar cell using amorphous titanium oxide as electron collection electrode inserted between ITO and organic layer, Sol. Energy Mater. Sol. Cells. 92 (2008) 1476– 1482. https://doi.org/10.1016/j.solmat.2008.06.012

[25]   T. Kuwabara, H. Sugiyama, T. Yamaguchi, K. Takahashi, Inverted type bulk-heterojunction organic solar cell using electrodeposited titanium oxide thin films as electron collector electrode, Thin Solid Films. 517 (2009) 3766–3769. https://doi.org/10.1016/j.tsf.2008.12.039

[26]   J.W. Kang, Y.J. Kang, S. Jung, M. Song, D.G. Kim, C.S. Kim, S.H .Kim, Fully spray-coated inverted organic solar cells, Sol. Energy Mater. Sol. Cells. 103 (2012) 76–79. https://doi.org/10.1016/j.solmat.2012.04.027

[27]   R.J. Peh, Y. Lu, F. Zhao, C.L.K. Lee, W.L. Kwan, Vacuum-free processed transparent inverted organic solar cells with spray-coated PEDOT:PSS anode, Sol. Energy Mater. Sol. Cells. 95 (2011) 3579–3584. https://doi.org/10.1016/j.solmat.2011.09.018

[28]   L. Mao, Q. Chen, Y. Li, Y. Li, J. Cai, W. Su, S. Bai, Y. Jin, H.Q. Ma, Z. Cui, L. Chen, Flexible silver grid/PEDOT:PSS hybrid electrodes for large area inverted polymer solar cells, Nano Energy. 10 (2014) 259–267. https://doi.org/10.1016/j.nanoen.2014.09.007

[29]   F.J. Lim, K. Ananthanarayanan, J. Luther, G.W. Ho, Influence of a novel fluorosurfactant modified PEDOT:PSS hole transport layer on the performance of inverted organic solar cells, J. Mater. Chem. 22 (2012) 25057-25064. https://doi.org/10.1039/C2JM35646E

[30]   B. Zimmermann, U. Wurfel, M. Niggemann, Long term stability of efficient inverted P3HT:PCBM solar cells, Sol. Energy Mater. Sol. Cells. 93 (2009) 491–496. https://doi.org/10.1016/j.solmat.2008.12.022

[31]   Y. Liu, C.C. Chen, Z. Hong, J. Gao, M. Yang, H. Zhou, L. Dou, G. Li, Y. Yang, Solution-processed small-molecule solar cells: breaking the 10% power conversion efficiency, Sci. Rep. 3 (2013) 3356. https://doi.org/10.1038/srep03356

[32]　D. Lee, W.K. Bae, I. Park, D.Y. Yoon, S. Lee, C. Lee, Transparent electrode with ZnO nanoparticles in tandem organic solar cells, Sol. Energy Mater. Sol. Cells. 95 (2011) 365–368. https://doi.org/10.1016/j.solmat.2010.04.020

[33]　M. Li, K. Gao, X. Wan, Q. Zhang, B. Kan, R. Xia, F. Liu, X. Yang, H. Feng, W. Ni, Y. Wang, J. Peng, H. Zhang, Z. Liang, H.L. Yip, X. Peng, Y. Cao, Y. Chen, Solution-processed organic tandem solar cells with power conversion efficiencies >12%, Nat. Photonics. 11 (2017) 85-90. https://doi.org/10.1038/nphoton.2016.240

[34]　D.J.D. Moet, P.D. Bruyn, P.W.M. Blom, High work function transparent middle electrode for organic tandem solar cells, Appl. Phys. Lett. 96 (2010) 153504. https://doi.org/10.1063/1.3387863

[35]　Y. Zhou, C.F.Hernandez, J.W. Shim, T.M. Khan, B. Kippelen, High performance polymeric charge recombination layer for organic tandem solar cells. Energy Environ. Sci. 5 (2012) 9827-9832. https://doi.org/10.1039/C2EE23294D

[36]　Z. Zhang, X. Zhang, H. Xu, Z. Liu, S. Pang, X. Zhou, S. Dong, X. Chen, G. Cui, CuInS$_2$ Nanocrystals/PEDOT:PSS composite counter electrode for dye-sensitized solar cells, ACS Appl. Mater. Interfaces. 4 (2012) 6242-6246. https://doi.org/10.1021/am3018338

[37]　T. Yohannes, O. Inganäs, Photoelectrochemical studies of the junction between poly[3-(4-octylphenyl)thiophene] and a redox polymer electrolyte, Sol. Energy Mater. Sol. Cells, 51 (1998) 193-202. https://doi.org/10.1016/S0927-0248(97)00213-4

[38]　Y.G. Tian, W.J. Huai, X.Y. Ming, L.J. Ming, H.M. Liang, F.L. Qing, Y. Ying, A dye-sensitized solar cell based on PEDOT:PSS counter electrode, Chin. Sci. Bull. 58 (2013) 559-566. https://doi.org/10.1007/s11434-012-5352-3

[39]　J. Ma, S. Yuan, S. Yang, H. Lu, Y. Li, Poly(3,4-ethylenedioxythiophene)/reduced graphene oxide composites as counter electrodes for high efficiency dye-sensitized solar cells, Appl. Surf. Sci. 440 (2018) 8–15. https://doi.org/10.1016/j.apsusc.2018.01.100

[40]　J. Xia, N. Masaki, K. Jiang, S. Yanagida, The influence of doping ions on poly(3,4-ethylenedioxythiophene) as a counter electrode of a dye-sensitized solar cell, J. Mater. Chem. 17 (2007) 2845–2850. https://doi.org/10.1039/B703062B

[41]　K. Kitamura, S. Shiratori, Layer-by-layer self-assembled mesoporous PEDOT–PSS and carbon black hybrid films for platinum free dye-sensitizedsolar-cell counter

electrodes, Nanotechnology. 22 (2011) 195703. https://doi.org/10.1088/0957-4484/22/19/195703

[42] G. Guan, Z. Yang, L. Qiu, X. Sun, Z. Zhang, J. Ren, H. Peng, Oriented PEDOT:PSS on aligned carbon nanotubes for efficient dye-sensitized solar cells, J. Mater. Chem. A. 1 (2013) 13268–13273. https://doi.org/10.1039/C3TA12669B

[43] C.P. Lee, C.A. Lin, T.C. Wei, M.L. Tsai, Y. Meng, C.T. Li, K.C. Ho, C.I. Wu, S.P. Laud, J.H. He, Economical low-light photovoltaics by using the Pt-free dye-sensitized solar cell with graphene dot/PEDOT:PSS counter electrodes, Nano energy. 18 (2015) 109-117. https://doi.org/10.1016/j.nanoen.2015.10.008

[44] C.T. Li, C.P. Lee, Y.Y. Li, M.H. Yeha, K.C. Ho, A composite film of TiS$_2$/PEDOT:PSS as the electrocatalyst for the counter electrode in dye-sensitized solar cells, J. Mater. Chem. A. 1 (2013) 14888–14896. https://doi.org/10.1039/C3TA12603J

[45] W. Maiaugree, S. Pimanpang, M. Towannang, S. Saekow, W. Jarernboon, V. Amornkitbamrung, Optimization of TiO$_2$ nanoparticle mixed PEDOT–PSS counter electrodes for high efficiency dye sensitized solar cell, J. Non-Cryst. Solids. 358 (2012) 2489–2495. https://doi.org/10.1016/j.jnoncrysol.2011.12.104

[46] T. Muto, M. Ikegami, K. Kobayashi, T. Miyasaka, Conductive polymer-based mesoscopic counter electrodes for plastic dye-sensitized solar cells, Chem. Lett. 36 (2007) 804-805. https://doi.org/10.1246/cl.2007.804

[47] H. Xu, X. Zhang, C. Zhang, Z. Liu, X. Zhou, S. Pang, X. Chen, S. Dong, Z. Zhang, L. Zhang, P. Han, X. Wang, G. Cui, Nanostructured titanium nitride/PEDOT:PSS composite films as counter electrodes of dye-sensitized solar cells, ACS Appl. Mater. Interfaces. 4 (2012) 1087–1092. https://doi.org/10.1021/am201720p

[48] M.H. Yeh, L.Y. Lin, C.P. Lee, H.Y. Wei, C.Y. Chen, C.G. Wu, R. Vittala, K.C. Ho, A composite catalytic film of PEDOT:PSS/TiN–NPs on a flexible counter-electrode substrate for a dye-sensitized solar cell, J. Mater. Chem. 21 (2011) 19021-19029. https://doi.org/10.1039/C1JM12428E

[49] D. Song, M. Li, Y. Li, X. Zhao, B. Jiang, Y. Jiang, Highly transparent and efficient counter electrode using SiO$_2$/ PEDOT–PSS composite for bifacial dye-sensitized solar cells, ACS Appl. Mater. Interfaces. 6 (2014) 7126-7132. https://doi.org/10.1021/am500082x

[50]   P. Sudhagar, S. Nagarajan, Y.G. Lee, D. Song, T. Son, W. Cho, M. Heo, K. Lee, J. Won, Yong S. Kang, Synergistic catalytic effect of a composite (CoS/PEDOT:PSS) counter electrode on triiodide reduction in dye-sensitized solar Cells, ACS Appl. Mater. Interfaces. 3 (2011) 1838-1843. https://doi.org/10.1021/am2003735

[51]   W. Maiaugree, A. Karaphun, A. Pimsawad, V. Amornkitbamrung, E. Swatsitang, Influence of $SrTi_{1-x}Co_xO_3$ NPs on electrocatalytic activity of $SrTi_{1-x}Co_xO_3$ NPs/PEDOT-PSS counter electrodes for high efficiency dye sensitized solar cells, Energy. 154 (2018) 182-189. https://doi.org/10.1016/j.energy.2018.04.122

[52]   S. Xu, Y. Luo, G. Liu, G. Qiao, W. Zhong, Z. Xiao, Y. Luo, H. Ou, Bifacial dye-sensitized solar cells using highly transparent PEDOT:PSS films as counter electrodes, Electrochim. Acta. 156 (2015) 20-28. https://doi.org/10.1016/j.electacta.2014.12.174

[53]   A.S.A. Ahmed, W. Xiang, X. Hu, C. Qi, I.S. Amiinu, X. Zhao, Si_3N_4/MoS_2-PEDOT:PSS composite counter electrode for bifacial dyesensitized solar cells, Sol. Energy. 173 (2018) 1135–1143. https://doi.org/10.1016/j.solener.2018.08.062

[54]   M. Pietsch, S. Jäckle, S. Christians, Interface investigation of planar hybrid n-Si/PEDOT:PSS solar cells with open circuit voltages up to 645 mV and efficiencies of 12.6 %, Appl. Phys. A. 115 (2014) 1109–1113. https://doi.org/10.1007/s00339-014-8405-4

[55]   J. P. Thomas, K. T. Leung, Mixed co-solvent engineering of PEDOT:PSS to enhance its conductivity and hybrid solar cell properties, J. Mater. Chem. A. 4 (2016) 17537-17542. https://doi.org/10.1039/C6TA07410C

[56]   L. He, Rusli, C. Jiang, H. Wang, D. Lai, Simple approach of fabricating high efficiency Si nanowire/conductive polymer hybrid solar cells, IEEE. Electron. Device. Lett. 32 (2011) 1406 – 1408. https://doi.org/10.1109/LED.2011.2162222

[57]   L. Hong, Rusli, X. Wang, H. Zheng, H. Wang, H. Yu, Design guideline of Si Nanohole/PEDOT:PSS hybrid structure for solar cell application, Nanotechnology. 24 (2013) 355301-355306. https://doi.org/10.1088/0957-4484/24/35/355301

[58]   S.A. Moiz, A.M. Nahhas, H.D. Um, S.W. Jee, H.K. Cho, S.W. Kim, J.H. Lee, A stamped PEDOT:PSS–silicon nanowire hybrid solar cell, Nanotechnology. 23 (2012) 145401. https://doi.org/10.1088/0957-4484/23/14/145401

[59]   K.T. Park, H.J. Kim, M.J. Park, J.H. Jeong, J. Lee, D.G. Choi, J.H. Lee, J.H. Choi, 13.2% efficiency Si nanowire/PEDOT:PSS hybrid solar cell using a transfer-

imprinted Au meshelectrode Sci. Rep. 5 (2015) 12093.
https://doi.org/10.1038/srep12093

[60]   H.J. Syu, S.C. Shiu, C.F. Lin, Silicon nanowire/organic hybrid solar cell with efficiency of 8.40%, Sol. Energy Mater. Sol. Cells. 98 (2012) 267–272. https://doi.org/10.1016/j.solmat.2011.11.003

[61]   M. Sharma, P.R. Pudasaini, F.R. Zepeda, D. Elam, A.A. Ayon, Ultrathin, flexible organic−inorganic hybrid solar cells based on silicon nanowires and PEDOT:PSS, ACS Appl. Mater. Interfaces. 6 (2014) 4356−4363. https://doi.org/10.1021/am500063w

[62]   M. Junghanns, J. Plentz, G. Andrä, A. Gawlik, I. Höger, F. Falk, PEDOT:PSS emitters on multicrystalline silicon thin-film absorbers for hybrid solar cells, Appl. Phys. Lett. 106 (2015) 083904. https://doi.org/10.1063/1.4913869

[63]   W. Lu, C. Wang, W. Yue, L. Chen, Si/PEDOT:PSS core/shell nanowire arrays for efficient hybrid solar cells, Nanoscale. 3 (2011) 3631-3634. https://doi.org/10.1039/C1NR10629E

[64]   K. A. Nagamatsu, S. Avasthi, J. Jhaveri, J. C. Sturm, A 12% efficient silicon/PEDOT:PSS heterojunction solar cell fabricated at < 100 ∘C, IEEE Journal of Photovoltaics. 4 (2014) 260 − 264. https://doi.org/10.1109/JPHOTOV.2013.2287758

[65]   J. Sheng, K. Fan, D. Wang, C. Han, J. Fang, P. Gao, J. Ye, Improvement of the SiO$_x$ passivation layer for high-efficiency Si/PEDOT:PSS heterojunction solar cells, ACS. Appl. Mater. Interfaces. 6 (2014) 16027−16034. https://doi.org/10.1021/am503949g

[66]   H. Wang, J. Wang, Rusli, Hybrid Si nanocones/PEDOT:PSS solar cell, Nanoscale. Res. Lett. 10 (2015) 191. https://doi.org/10.1186/s11671-015-0891-6

[67]   D. Zielke, A. Pazidis, F. Werner, J. Schmidt, Organic-silicon heterojunctionsolarcellson n-type siliconwafers: The back PEDOT concept, Sol. Energy Mater. Sol. Cells. 131 (2014) 110-116. https://doi.org/10.1016/j.solmat.2014.05.022

[68]   S. Li, Z. Pei, F. Zhou, Y. Liu, H. Hu, S. Ji, C. Ye, Flexible Si/PEDOT:PSS hybrid solar cells, Nano. Res. 8 (2015) 3141-3149. https://doi.org/10.1007/s12274-015-0814-y

Materials for Solar Cell Technologies II
Materials Research Foundations **104** (2021) 77-113

Materials Research Forum LLC
https://doi.org/10.21741/9781644901410-4

Chapter 4

Transparent Conducting Electrodes for Optoelectronic Devices: State-of-the-art and Perspectives

Abhijit Ray[1*], Rajaram Narasimman[2]

[1]Department of Solar Energy, and Solar Research and Development Center, Pandit Deendayal Energy University, Gandhinagar – 382426, Gujarat, India

[2]Advanced Power Systems division, Chemicals System Group, Vikram Sarabhai Space Centre, Thiruvananthapuram-695022, Kerala, India

*abhijit.ray1974@gmail.com; abhijit.ray@sse.pdpu.ac.in

Abstract

This chapter brings a concise review of the transparent conducting materials, films and electrodes (TCM, TCF and TCE, respectively), its state-of-the-art and outlooks ahead. Initial part of the chapter gives a general introduction of the topic, followed by a feasible road map as proposed and collated by the authors based on several other reviews. Fundamental physics behind the transparent conductors is discussed in the latter part. Established and potential oxide based TCMs, namely the transparent conducting oxides (TCOs) are reviewed which are being used commercially and will see application in the near future. Non-conventional TCMs, which are mostly non-TCOs, such as graphene, carbon nanotubes (CNT), metallic nanowires (MNWs) and their hybrids are described in brief. Scalability and large area fabrication which are most important concerns for commercialization of TCMs are discussed. The general prospects are given at the end.

Keywords

Transparent Conductors, Thin Films, Metallic Nanowires, Graphene

Contents

1. Introduction

One of the important components of the optoelectronic devices like touch screens [1,2], flat-panel displays [3,4], light-emitting diode (LED) [5], solar cells [6], photosensors [7] and e-papers [8] are transparent conducting electrode (TCE). TCEs are transmitting the light and conducting the electrical current, simultaneously, mostly in the visible region. Flexible, bendable, stretchable transparent electronic devices have gained a huge interest in the academic as well as industrial research in the recent years.

The first example of TCE materials was thin metal films prepared by evaporating and sputtering reported in the end of the nineteenth century [9]. Baedeker [10] reported the development of CdO, Cu_2O and PbO in 1907 which are the early examples of the transparent conducting oxides (TCOs) [11]. The resistivity of CdO was 1.2×10^{-3} Ω with a yellowish appearance. This resistivity is around one order less than the ITO films. Currently, indium tin oxide (ITO), a transparent conducting oxide (TCO), is dominating the TCEs for almost four decades due to its excellent conductivity (10 Ω sq^{-1} on glass) and high optical transmittance (>90%) [11]. Growing technologies and demand in the display industries for the flexible displays demands transparent conducting electrodes with excellent flexibility which is retaining its transmittance and electrical conductivity [12]. However, ITO has few serious drawbacks *viz*, poor flexibility and brittleness because ITO is a ceramic material [13]. Also, the cost of indium is increasing rapidly due to its low-availability in the earth's crust; however, synthetic procedure of the ITO is more expensive. Other oxide-based transparent electrodes such as $SnO_2{:}F$ (FTO) [14] and ZnO:Al (AZO) [15] are also studied due to low-cost when compared to the ITO. However, they are also having the same limitation as of ITOs; like vapour phase synthesis and poor flexibility. This demands the suitable replacement for the conventional ITO with new materials with good transmittance and electrical conductivity along with excellent flexibility for the future devices [16].

Poly (3,4-ethylenedioxythiophene):poly(styrene-sulfonate) (PEDOT:PSS) is a conducting polymer which has been widely studied for the TCEs. PEDOT:PSS is also used in organic photovoltaic and organic light emitting diode as the hole conducting layer [17]. It has huge advantage such as low-cost, transparent, flexible and solution-processable for the large scale fabrication of conducting polymer based-TCEs. However, the conducting polymers have the disadvantage of low specific conductivities when compared to ITO. Nanocarbons such as graphene [18-20] and carbon nanotubes (CNTs) [21-23] are extensively studied for the TCE applications due to excellent electrical conductivity and optical transparency. Especially, single-walled carbon nanotube (SWCNT) films and single-layer graphene (SLG) have attracted significant research interest due to such properties [24,25].

One-dimensional metal nanostructures gained significant importance because of its high aspect ratio, ability to form conducting network and good crystallinity [26-28]. When compared to the CNT network films, 1D metal nanostructures show good contact between the nanowires whereas CNT shows high contact resistance in tube-tube junction [29,30]. Silver (Ag) [31-35] and copper (Cu) nanowires (NWs) [1, 36-39] networks are widely studied for the TCEs. The advantages of the metal nanowires are solution-processable, high optical transparency, low sheet resistance and more importantly, compatible with the flexible substrates.

The materials studied for the TCEs are diverse and exhibit spectrum of properties that find different applications in displays, heaters and sensors. In this chapter, critical description of the recent research progress in the field of TCEs with emerging new materials, which have the potential replacement for the conventional ITO, the technological roadmap for the TCEs in the commercial markets , the physical aspects of the materials used in the TCEs, overview of oxide-based TCEs, focuses on the hybrids of materials used for the TCEs and the applications of these materials in various optoelectronic devices with a comparison with ITO is given.

2.   Roadmap to the transparent electrode technologies

The market for transparent conductive films and materials (TCF and TCM) has been changing at a rapid pace. Such a transformation is taking place at all levels of the value chain, including the technologies, applications and supplies [40]. New materials and technologies beyond expensive (indium doped Tin Oxide) ITO are imperative, for example, metallic nanowire (MNW), such as, Ag-nanowires and nanoparticels, nano-carbon (graphene and carbon nanotubes) and conductive polymer (PEDOT etc based transparent conductors. These new materials should be compatible with applications such as, add-on and embedded touch technologies in mobiles, smart watches, tablets, notebooks, all-in-one display-based gadgets (AiOs), automotive displays, smart windows, organic LEDs (OLEDs), organic photovoltaics (OPVs) and other thin film photovoltaics, in-mold electronics (IME), OLED lighting, transparent heating, and so on. Among these fixed as well as flexible OLED display markets are in immediate forefront in the recent decade. A major technology shift has been observed since 2008 when touchscreen-based devices and gadgets started invading the commercial sectors and consumer durables. In more recent years, projected capacitive touch screen (PCAP) technologies demand TCMs with higher mechanical flexibilities without compromising the electrical conductivity and optical transparency. Not just the PCAP, but the rise of embedded touch screens in various microelectronic products and systems, which needs TCF to be stable and more durable under operations in stringent environmental conditions, like high (>40 °C) or low (<0 °C) ambient temperature, shear and vibrations etc.

In view of the above considerations, the roadmap of the TCF and TCM technologies should be set based on the fundamental materials properties, their realization and the market trends as illustrated in Figure 1. In following subsections, each one has been elaborated.

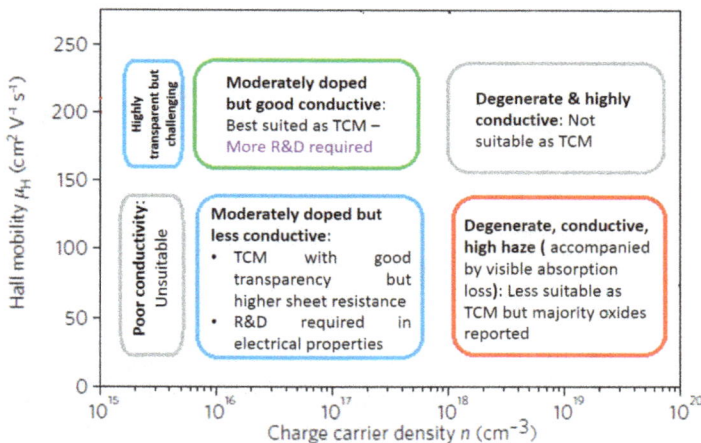

Figure 1 *Representation of competing electrical mobility vs. free carrier concentration in TCO materials. The red marked region is where most of the TCOs as on date falls. The blue marked zones are where R&Ds are ongoing as transmittance (T) of visible photons is good for the materials falling in them. Some challenges exist as well in terms of materials design and process feasibility. The green marked region is where more R&D is required as obtaining a large mobility at lower carrier concentration will satisfy both high T and electrical conductivity.*

2.1   Roadmap towards improvement of material properties

In general, the transmission cut-off in a TCF/TCM is determined in short-wavelength (UV) and long wavelength (IR) regions, by intrinsic band gap absorption and the

Materials for Solar Cell Technologies II   Materials Research Forum LLC
Materials Research Foundations **104** (2021) 77-113   https://doi.org/10.21741/9781644901410-4

absorption onset by concentration dependent conduction band electron plasma oscillation, respectively. The figure of merit of the TCM may be determined by high electrical conductivity (σ) and optical transmittance (T) in visible region of spectrum. A higher T is warranted in different TCMs through different mechanisms. For example, in oxide based TCMs, such as SnO_2, ZnO, CdO, In_2O_3etc, high T is obtained through wide band gap (>3 eV) without the presence of colour centres (defect states within gap). In MNW based TCMs, the spatial density of nanowires is to be critically optimized. The high visible photon transmission also requires that the plasma resonance frequency should be large to shift absorption peak towards IR photon energies. The wavelength corresponding to the plasma resonance frequency (λ_p) is a function of the free carrier concentration (n) as,

$$\lambda_p \sim \frac{1}{\sqrt{n}} \tag{2.1}$$

As the lower magnitude of absorption is desirable in the visible photon energy, the shift of λ_p at longer wavelength demands the TCMs are provided low doping concentration. This however will compromise the electrical conductivity (σ) of the semiconductor through the standard expression:

$$\sigma = q\left(n\mu_n + p\mu_p\right) \tag{2.2}$$

where, q is the electron charge, n/p is free electron/ hole concentration, and μ_n/μ_p is the electron/hole mobility. Therefore, in order to enhance σ, the carrier mobility should be enhanced which can be done by modifying the band structure of host crystal near the band edges.

2.2 Roadmap towards new technology realization

2.2.1 p-type TCOs

At present, a great majority of TCMs available in market are n-type semiconductor based transparent conducting oxides (TCOs) because of their easy dupability to achieve a long range of doping (*e.g.* from 10^{14}-10^{19} cm^{-3}). This doping is usually done through substitution at host compound through aliovalent species or hetero-atoms both via solid state or solution processes. Another reason behind the popularity of n-type TCOs is that conduction band of most TCO crystals are populated by delocalized s-orbitals, which ensures a very high electron mobility. On the other hand, p-type TCOs are suffering from low conductivity originating from large effective mass of holes leading to low hole mobility in most oxide materials [41]. This is one of the great hindrances in the commercialization of p-type TCO materials. The valence band maximum in the wide band gap oxides are predominantly occupied by localized oxygen p orbitals, which causes deep acceptor levels and hence a poor hole mobility. Therefore, in order to promote p-type TCM technologies, a number of fundamental aspects need to be thoroughly addressed in further research [42]:

1. Create shallower acceptor levels by reducing defect ionization energy through transition or rare earth metal doping, co-doping, multivalent impurity doping etc.

2. Increase defect solubility through non-equilibrium synthesis method, such as hydrothermal, solvothermal, non-stoichiometric routes, or through increasing host energy by surfactants.

3. Engineer the band structure to achieve higher hole mobility by reducing the hole-effective masses. Increasing the doping concentration at degeneracy level needs to be controlled to reduce visible photon absorption as well as its reflections.

2.2.2 Metallic nanowire based TCMs

Metals with low bulk resistivity, such as Al, Ag and Cu in their nano-mesh forms are promising TCMs. A great number of developments have been seen in past decade over Ag- and Cu- based nanowires and patterns [28]. However, due to their high free carrier concentration $\sim 10^{22}$-10^{23} cm^{-3}, the corresponding plasma wavelengths lie in visible to deep ultraviolet regime and it is therefore, challenging to make them highly transparent in the visible photon energies. Owing to this, the metallic films would be highly reflective in the visible region and have a large absorbance in the UV region. However, many such metal films and their nano-mesh are highly transparent in the IR region by virtue of the above consideration of blue shifted plasma wavelength. Therefore, the nano-metallic films would be an interesting option in IR applications, such as IR-sensors, IR-photo detectors and IR-solar cells (or thermophotovoltaic devices).

Optoelectronic applications of metallic TCMs are promising as transparency of the TCFs can be made as high as 85% or above by optimizing the size (length, diameter etc) and shape of the metallic nanostructure. However, their sheet resistance (Rs) has been found to increase in reduced dimension due to intrinsic grain boundary scattering of free carriers. Beside their physical aspect, chemical stability is another important factor for the development of metallic TCMs. For example, oxidation resistance of Ag is good and therefore Ag-nanowires need not have much protection from ambient. However, oxidation probability increases from Ag to Cu to Al (by virtue of their position in the electrochemical series). In nano-metallic TCFs therefore, few important aspects should be well addressed before they can be integrated into commercial optoelectronics: 1. Cost effectiveness at large scale, 2. Transparency in visible region, 3. Sheet resistance and 4. Oxidation resistance. In table 1 [27,39,43,44] a scope is defined for their future development based on above four aspects and the untapped applications as on date.

Table 1 A roadmap towards metal nanowire based TCE

Metal option	Timeline	Transparency in visible region (Transmittance)	Sheet resistance Rs $(\Omega/\square)$	Cost effectiveness at large scale	Oxidation probability	Application target
Silver (Ag)	To date	~ 80% [43]	~ 38 $\Omega/\square$	Not feasible	Low	Thin-film, Organic PV and LEDs
	Future target	• Low cost processing • Immediate deployment in touch screen technologies				
Copper (Cu)	To date	~ 85% [27,39]	~ 60 $\Omega/\square$	Highly cost effective	Moderate	Organic PV and OLED
	Future target	• Reduction in Rs through nano engineering. • Improve oxidation resistance • Application over thin film PV modules (replacing ITO)				
Aluminium (Al)	To date	~ 80% [44]	~ 45 $\Omega/\square$	Highly cost effective	Very high at nanoscale	
	Future target	No much promise due to low mobility and chemical stability.				

2.2.3 Carbon based TCMs

There are two major carbon based TCMs studied and developed so far: carbon nanotubes (CNTs) and graphene. The conducting polymers, such as PEDOT:PSS may be included in this category, however, such polymeric TCFs suffer from poor sheet resistance-transmittance trade-off, as well as chemical and thermal instability of their electrical conductivity limiting these types of TCMs in sustainable applications [45]. Amid high interflake resistances in graphene, the graphene as TCM holds some remarkable advantages over the ITOs in especially organic photovoltaic (OPV) and organic LED (OLED) devices. Novoselov et al. [46] has provided an excellent roadmap for the use of graphene in next 15 years in his review. The review anticipates a transition from medium quality graphene in touch screens, e-papers and foldable OLEDs as of now to high quality large area graphene in future developments. The graphene is expected to finally see a market in composited forms, such as that with CNTs, polymer matrices etc.

Presently, graphene has already been attempted as TCMs in optoelectronic technologies, such as flexible-OLEDs, touch screens and e-papers. However, it is only partial due to the high cost involvement in the chemical vapor deposition (CVD) process involved in the graphene fabrications. Therefore, dependency on ITO has not been fully eliminated. In the next 10 years (by 2030), touch screens will be required to be interactive, such as car windscreens, industrial process and control view interfaces etc. Besides, read-write on e-paper will be done on foldable interface with radius as small as 5mm. The high contact resistance during folding has been a challenge which is required to be resolved. Graphene based TCFs will also be required on flexible solar panels by that time. In the next 20 years (by 2040) above technologies in wearable forms will be significant. They are also expected to be interactive and internet of things (IoT) based. As these wearables will be used in various weather conditions, such as high/low temperature/ humidity, their structural and thermal stability are required to be investigated in detail. In the next 30 years, the graphene based TCMs are expected to invade biotechnology where embedded soft electrodes can handle various functionalities of vision such as photochromic, electrochromic etc. implants in human eyes (Figure 2).

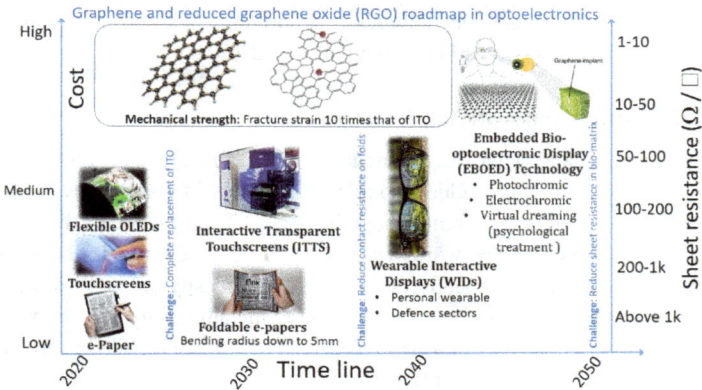

Figure 2 A 30-years roadmap for graphene and reduced graphene oxide based TCMs in optoelectronic applications.

2.3   Market trends of transparent conductors

The market of transparent conductive films is expected to reap $5.86 billion by 2020 according to a private market survey [47], with a cumulative annual growth rate (CAGR) of 17.2% during the last forecast period of 2014 - 2020. The largest use of TCF as discussed above in display technologies is either on glass or on films. The market is now growing primarily due to an escalating demand for touch enabled devices. Declining cost of new smartphones, continuous adoption of touch-based user interfaces (self-service kiosks *etc*), energy efficiency, light weight, low-cost materials, flexibility and robustness are the key drivers of this market. Apart from this, OLED lighting, OPV and DSSSC are also potential markets. However, these are expected to register a slower growth initially, but shall escalate in coming decade.

The TCF market is also segmented on the basis of technology, application and demographic acceptability. Key technologies which are implemented in manufacturing include ITO on PET, ITO on glass, non-ITO oxides, silver nanowires, graphene, carbon nanotubes, metal mesh, micro fine wire, and PEDOT [47]. The application industries include smartphones, tablets, notebooks, all-in-one PCs, monitors, TV displays, and OLED lighting. The geographical demand goes to North America, Europe, Asia Pacific (AP) and Latin America, Middle East and Africa (LAMEA). Among these, AP has been the leading region in the TCF market followed by North America. As per the above, the key companies in this market include (not limited to) 3M, Nitto Denko Corp, Toyobo Corp, Dupont Teijin Films, Eastman Kodak, Fujifilm Holdings Corp, Canatu Oy, DONTECH Inc, Cambrios Technologies Corporation, and Rolith. Various other companies are also growing in China, South Korea and Taiwan.

3.   Physical aspects and materials for TCEs

The most common and vital requirements for a TCE are high conductivity and transparency. Typically, semiconductor oxide films with wide band gaps are best suited for its transparent nature (transparent conducting oxide, TCO). Besides, there can also be metallic mesh or nano-carbon networks which may be considered as TCEs. Transparency in the later class materials depends on their spatial density in most cases. As their material properties are totally different from TCO to metallic or nano-carbon TCEs, their electrical conductivity needs to be reviewed in different manner. These three classes of TCEs with respect to their physical properties are discussed below.

3.1   Physical requirement and materials for TCOs

In semiconducting oxide (TCO) materials a high conductivity σ can be warranted through a high carrier concentration n (electrons) of p (holes) and/or a sufficiently high carrier mobility $\mu_{n,p}$ according to following equation:

$$\sigma = q(n\mu_n + p\mu_p) \tag{3.1}$$

Materials for Solar Cell Technologies II   Materials Research Forum LLC
Materials Research Foundations **104** (2021) 77-113   https://doi.org/10.21741/9781644901410-4

where, q is the electron charge.

By virtue of small effective mass, electrons have larger mobility than that of holes, and therefore, n-type TCOs are usually preferred over the p-type. However, the carrier concentration cannot be taken as a design parameter to improve the electrical conductivity because it is limited by the absorption of light by free electron gas, which compromises the visible light transmittance. According to classical Drude theory electrical field of a light wave causes the collective plasma excitation in electrons and defines a plasma frequency given by [48]:

$$\omega_p = \sqrt{\frac{ne^2}{m^* \varepsilon_r \varepsilon_0}}$$

(3.2)

where m* is the effective mass of the electron carriers and $\varepsilon_r \varepsilon_0$ is the permittivity of the material. In most of the TCOs (Table 2) therefore, it is essential to keep the free carrier concentration below few 10^{21} cm^{-3} to keep the plasma wavelength in the infra-red region. Therefore, the higher electrical conductivity must be assured through the carrier mobility in these materials. Therefore, the electronic band structure plays vital role to control the effective mass of their free carriers. Table 2 below summarizes some common TCO materials with their reported physical properties [29].

3.2 Physical requirement and materials for metallic TCEs

In metallic TCEs nanostructures such as metallic nanowires or nanomesh are adopted to ensure large transmittance values, electrical conductivity needs to be critically evaluated due to two reasons: (1) metals have already very large free carrier concentration (n~10^{23} cm^{-3}), which makes them to absorb visible light due to their low values of plasma wavelength cut-off, and (2) poor interconnects at their nano-domain.

3.3 Physical requirement and materials for nano-carbon TCEs

Among various nano-carbon variants, graphene has attracted highest attraction in the field of TCE due to its outstanding two-dimensional electron gas at room temperature. Graphene shows exclusive electronic band structure where the valence band is formed by bonding π-orbitals, whereas the conduction band is formed by the anti-bonding π*-orbitals. The conduction band eventually meets the valence band at a single point crystal momentum, $k = 0$, commonly known as a Dirac point [49]. At this point, the electron energy (E) is linearly dependent on k. This linear dispersion relation results in massless excitons, and thus electrons in graphene behave like massless Dirac fermions which can travel large distances (in μm to tens of μm) as compared to their domain size (typically 10-100s nm) without scattering (ballistic transport). This leads to very high electron mobility at low carrier concentrations (table 2- indicative values). This intriguing property of graphene shifts the plasma wavelength to the far-infrared and it becomes highly transparent to visible light. Although single layer graphene is characterized with high intrinsic mobility $\sim$ 5000 cm^2/V.s, large area TCFs using graphene face real challenge of the presence of wrinkles, cracks, few-to-multi layers and other mobility-reducing defects and therefore, are yet to be addressed.

Table 2 *Parameters (reported so far) of various possible TCMs which determine its application merit.*

TCM example	Bulk resistivity (Ω-cm)	Free electron concentration (cm^{-3})	Plasma wavelength (nm)	Electron mobility (cm^2/V.s)	Merit (M)/ Demerit (D)
Semiconducting oxides					
ITO (In:SnO$_2$)	$\geq 10^{-4}$	$< 3 \times 10^{21}$	770	$\sim$ 20-100	M: High visible transmittance, very large electrical conductivity, low roughness. D: Highly brittle (low mechanical strength)
FTO (F: SnO$_2$)	$\geq 5 \times 10^{-4}$	$< 10^{21}$	1100	$\sim$ 15-50	M: (Close to ITO) D: High roughness, higher processing temperature than ITO — application limitations.
AZO (Al 2%:ZnO)	$\geq 3 \times 10^{-3}$	$< 3 \times 10^{20}$	1100	$\sim$ 20	M: Cheaper than ITO & FTO due to ZnO D: RF-sputtering only best reported technique for TCF (process limitation)
BZO (B$_2$O$_3$ 2%: ZnO)	$\geq 2 \times 10^{-2}$	$< 10^{20}$	1100	$\sim$ 2-5	M: Visible transmittance better than AZO D: Low electrical conductivity due to low mobility as well as carrier concentration.
Metals based					
Cu	$\sim 2 \times 10^{-6}$	$\sim 8.5 \times 10^{22}$	110	$\sim$ 50	M: Highly conductive, cheap D: Prone to oxidation, visible light absorption (high haze)
Ag	$\sim 1.5 \times 10^{-6}$	$\sim 6 \times 10^{22}$	140	$\sim$ 72	M: Very large conductivity, oxidation resistant. D: Cost, high haze.
Al	$\sim 2.5 \times 10^{-6}$	$\sim 1.8 \times 10^{23}$	80	$\sim$ 14	M: Cheap D: Very high visible absorption and haze factor.
Carbon based					
Graphene	$> 5 \times 10^{-6}$	$< 3 \times 10^{20}$	1900	< 5000	M: Extremely large electrical conductivity in single layer 2D form, highly transparent. D: Scaling up not easy.
SW-CNT	$> 1.5 \times 10^{-4}$	$\sim 3 \times 10^{17}$	250	< 10	M: Cheap and ductile. D: High haze above certain concentration / surface coverage.
Conductive polymer	$> 7 \times 10^{-4}$	$\sim 10^{20}$	800	< 1	M: Highly ductile (best suited as flexible TCM) D: Sheet resistance-transmittance trade-off

Materials for Solar Cell Technologies II   Materials Research Forum LLC
Materials Research Foundations **104** (2021) 77-113   https://doi.org/10.21741/9781644901410-4

4.   Oxide based TCMs: The TCOs

This section provides a brief account of TCOs from fundamental material in relation to the viewpoint of physics and industrial applications. A brief review of two well-known TCOs, namely the ZnO and In_2O_3, and their doped forms are given. Their crystal and electronic structures are also discussed. The research challenges in the field of transparent electronics and solar energy are discussed in specific. In later part, the developing field of hole transporting TCOs (or p-type TCOs) are also accounted.

4.1   Physical and chemical basis of TCO materials

The TCOs are class of heavy metal oxides, which are wide band gap semiconductor by virtue of low-lying O-2p orbitals which mostly constitute the top of valence bands (VB). Their electrical conductivity (σ) is determined by the product of carrier mobility ($\mu_{n/p}$) and their concentration (n/p) along with the fundamental charge (1.6×10^{-19} C). Fundamentally, large σ may be warranted by either large n/p (degenerate semiconductor) or a large $\mu_{n/p}$. However, as will be discussed later degeneracy in TCEs are not advisable to prevent visible light absorption by plasmon resonance phenomenon. A best strategy is therefore, selecting oxides that will provide large $\mu_{n/p}$. The electron mobility (n-type semiconductor) is proportional to the width of conduction band (CB) and therefore, a large overlap between electron sharing orbital is required. A number of metal oxides, containing of a heavy metal cation with an outermost shell electronic configuration of (n-1) $d^{10}ns^0$ ($n \geq 4$) and oxygen anions satisfy this condition (Figure 3) [50].

As mentioned above, the top of the VB is constituted by the O-$2p$ orbitals and the bottom of the CB is mostly populated by the metal-ns (M-ns) orbitals (preferably to remain unoccupied) as illustrated in Figure4 for ITO (Sn doped In_2O_3). The spatial spreading of these M-ns is large and the overlap between them with spherical symmetry is also large. Metals with larger overlap integral between their n-s orbital offer higher mobility. Table 3 [51] below summarizes some common TCO metal candidates with their s-orbital overlap integrals. As is clear from table 3, s-orbital overlap is large in case of Zn and In, in ZnO and In_2O_3, respectively. Therefore, these two oxides are offering very high carrier mobility. SnO_2 also offers reasonably good electrical properties. Remaining candidates are either containing scare (Ga/Ge) or toxic (Cd) elements.

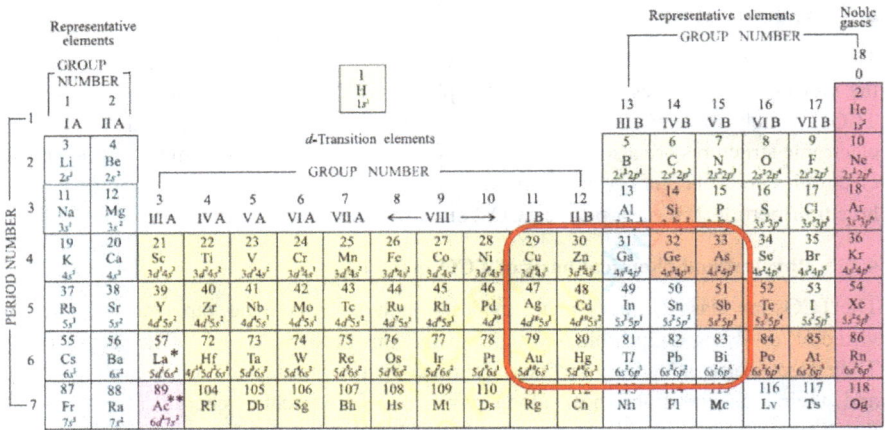

Figure 3 *Heavy metal cations as suitable candidates for TCOs. Periodic table ref [47].*

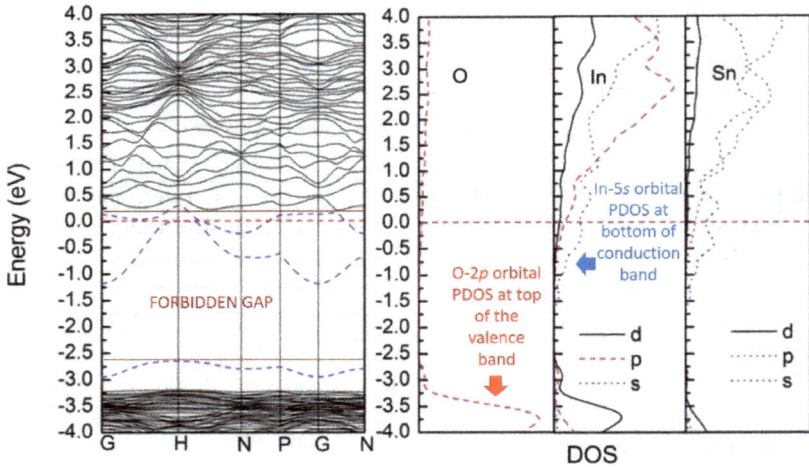

Figure 4 *Typical band structure and projected density of states (PDOS) of an ideal TCO material (ITO here).*

Table 3 *Metal conduction band orbital overlap integral values in some representative metal candidate for TCOs according to ref [51].*

TCO example	Orbital overlap	Overlap integral value	References
ZnO	Zn-4s/Zn-4s	0.604	[51]
In_2O_3	In-5s/In-5s	0.561	[51]
β-Ga_2O_3	Ga-4s/Ga-4s	0.463	[51]
GeO_2	Ge-4s/Ge-4s	0.284	[51]
CdO	Cd-5s/Cd-5s	0.69	[51]
SnO_2	Sn-5s/Sn-5s	0.452	[51]

4.2 Overview of common TCOs

As discussed above, both In_2O_3 and ZnO are non-toxic and have relatively large overlap integrals, which lead to a high electron mobility ensuring high electrical conductivity. In_2O_3 has a bixbyite cubic crystal structure which is a vacancy-defect oxide. The bixbyite structure is similar to that of fluorite and has a face-centred cubic array of indium atoms with all the tetrahedral interstitial positions filled with oxygen atoms. The MO_8 (metal centred- cornered oxygen) coordination units are replaced by missing oxygen from either the body or the face diagonal as depicted in Figure 5. The removal of two oxygen atoms from the MO_8 to form the MO_6 coordination units forces indiaum from the centre.

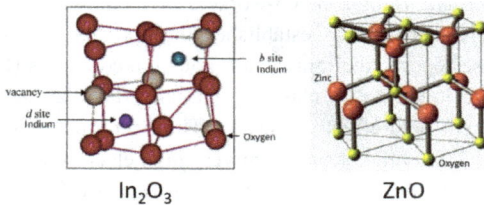

In_2O_3 ZnO

Figure 5 *Crystal structure of In_2O_3 and ZnO which are basis of common TCOs.*

In the In_2O_3, O^{2-} ions occupy $3/4^{th}$ of the tetrahedral interstices of an fcc In^{3+} array. Consequently, valence band of In_2O_3 consists of filled oxygen-2p states. The indium-3d core lies below the valence band edge. On the other hand, the conduction band is characterized by indium-5s band maintaining a gap of 3.75 eV with valence band. At high concentration of oxygen, an oxygen vacancy band forms and overlaps at the bottom of the conduction band. It makes In_2O_3 a degenerated semiconductor. In this case, the oxygen vacancies act as doubly ionized donors and contribute a maximum of two electrons to the conduction band. Additional improvement of conductivity can be done via doping Sn into In sites. Since indium exists as In^{3-}, the Sn^{4-} substitution results in an n-type doping by providing an extra electron to the conduction band in order to preserve the overall charge neutrality.

The second TCM of interest is ZnO, which crystallizes in a highly stable wurtzite structure as shown in Figure 5. The electronic structure of ZnO has been widely studied by many researchers [52]. In ZnO, the lowest two valence bands correspond to the O^{2-} : $2s$ core-like states and the next six valence bands are populated by O^{2-} : $2p$ bonding states. On the other hand, first two conduction bands (lowest of which is 3.4 eV apart from the valence band) are strongly localized by unoccupied Zn^{2+}: $4s$ levels to support electronic conduction. The undoped ZnO is slightly conducting, whereas higher conductivity occurs from the self-doping from the oxygen vacancies and/or Zn-interstitials. The donor levels can also be produced by the incorporation of foreign atoms such as H, In, Ga, B etc.

4.3   Emerging TCO substitutes of ITO

In the past few decades there have been only incremental developments with respect to basic n-type crystalline TCO properties and such a class of materials has almost reached a maturity. Hosono and co-workers [53] first pointed out that, amorphous oxides can be promising TCOs. Later, Narushima and co-workers [54], Nomura et al. [55], Hosono [56] and Taylor et al. [57] independently established the potential of amorphous oxides as TCOs [54-57]. In these reports, the ionic amorphous mixed oxides $(In_2O_3)_x(ZnO)_{1-x}$ and $(Ga_2O_3)_x(In_2O_3)_{1-x}$ constituted a new class of TCO materials that could be used both as TCEs and for transparent field-effect transistors. These amorphous semiconductors have unique electrical conductivity that stems from very high electron mobility ($\sim$10 cm^2 /V.s) despite large structural disorders. Such extraordinary electrical property exists because their conduction band is composed of spherical, isotropic 5s orbitals of the metal cation, the overlap of which does not change greatly in their amorphous state. Another promising new TCO material is doped TiO_2, which was first invented as TCE by Hasegawa and co-workers [58] in 2005. The resistivity reported in this system was remarkably low $\sim 5 \times 10^{-4}$ Ω cm. Although, this could be reached only at relatively high substrate temperatures of more than 300 °C, which is a drawback in many practical devices. A big advantage,

Materials for Solar Cell Technologies II   Materials Research Forum LLC
Materials Research Foundations **104** (2021) 77-113   https://doi.org/10.21741/9781644901410-4

however, with doped TiO_2 is its excellent chemical stability and further research is necessary to make it colourless as well.

4.4   Hole transporting or p-type TCOs – Challenges and perspectives

The transparent conducting oxides are crucial to many technologies from transparent electronics to thin film solar cells. While majority of the transparent conducting oxides in commercial sectors as mentioned above are n-type, their p-type counterparts are not well commercialized, as they exhibit much lower carrier mobilities that originates from large effective masses of hole in most oxides. Therefore, a primary objective to develop p-type TCO is to identify crystal structures which will offer low hole effective masses without compromising their wide band gap and low free carrier concentration. A fundamental difficulty in developing high-mobility p-type TCOs lies in the valence band in most oxides which is mostly populated by localized oxygen $2p$-orbitals. This makes the valence band quite flat across the Brillouin zones and it leads to a large hole effective masses ($m_h{}^*$) [59]. Kawazoe et al. [60] in 1997 first demonstrated that, delafossite compound and a p-type semiconductor, $CuAlO_2$ could show encouraging electrical conductivity and high optical transparency in visible light. In the crystal a large hole mobility was explained by a large hybridization of the oxygen orbitals with $3d^{10}$ electrons of Cu^{+1} p-closed shell, which lowers the oxygen's localized character and ultimately leads to dispersive valence band. Since then, the presence of Cu in p-type oxide semiconductor compounds has been believed to be a determining factor towards TCO development. In 2012, Hautier et al. [41] presented a design principle to identify the best suited candidates for p-type TCOs based on first-principles calculations. According to this design, several oxides can show p-type conductivity with wide band gap (> 3eV) but with effective masses ($0 < m_h{}^* < 1.5$) as compared to existing p-type TCOs, including $CuAlO_2$ (where $m_h{}^* \sim 2.5$ or above). This provides an opportunity to make them competitive with their n-type counter parts where the electron effective mass ranges between 0 and 0.5 (Figure 6- the target region of development). Apart from the copper d-orbital, hybridization in existing Cu-based p-type TCOs, there are also a number of cations, such as Sn^{2+}, Pb^{2+}, Tl^{1+} etc. where hybridization of s-states from $(n-1) d^{10}ns^2$ ($n \geq 5$) is possible with the oxygen-$2p$ states, making its effect benign. As the s-orbitals are more delocalized than d-orbitals, a higher dispersion of the valence band is possible to ensure lower effective masses of hole. Therefore, wide band gap compounds, such as $K_2Sn_2O_3$, $Rb_2Sn_2O_3$, $PbTiO_3$, $PbZrO_3$, $PbHfO_3$, $K_2Pb_2O_3$, Tl_4O_3 and $Tl_4V_2O_7$ are quire promising candidate as p-type TCOs provided feasible synthetic routes and their environmental stabilities (especially for Pb and Hg-based candidates) are ensured.

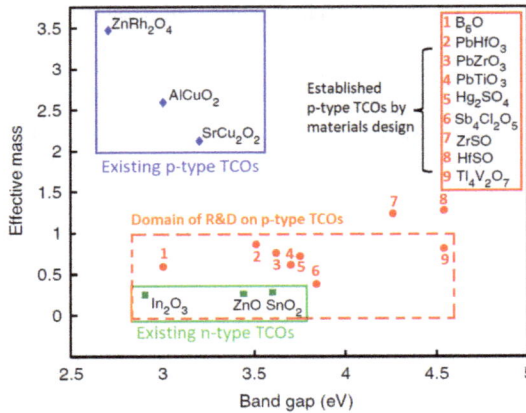

Figure 6 *Existing state of the art of p-type TCOs and their opportunity region of development in comparison to the existing ideal n-type TCOs.*

5. Nano-carbon based TCMs: Graphene and CNTs

5.1 Graphene based TCEs

Graphene is a two-dimensional nanomaterial where carbon atoms are arranged hexagonally planar one-atom thick layer. Graphene has been studied theoretically in the second half of the 20th century. In 2004, graphene was successfully synthesized by mechanical cleavage method from graphite. After that, graphene has been successfully prepared by various techniques and studied for different applications. Even though various preparation techniques are available, for graphene based TCEs are generally prepared from CVD grown graphene or from the graphene oxide (GO) [61]. GO can be directly coated over the substrate and can be reduced. In case of CVD grown graphene, the graphene has to be transferred into the device substrate which involves etching and stamping, thermal release method, photoresist method, roll-to-roll method and general method.

$$T = \left(1 + \frac{Z_0}{2R_S}\frac{G_0}{\sigma_{2D}}\right)^{-2}$$

where $Z_0 = 1/\varepsilon_0 c$ is the free-space impedance, ε_0 is the free-space dielectric constant, and c is the speed of light. For graphene sheet, $\sigma_{2D} = n\mu e$, where n is the number of charge carriers per unit volume, e is the electronic charge and μ is the mobility.

Graphene based TCEs were reported based on the solution processed GO for different photovoltaic devices. Wang et al. [62] showed that the GO prepared from the Hummer's method was deposited over the quartz substrate by dip coating and reduced by thermal treatment. The electrical conductivity of the graphene film was 550 S/cm and optical transmittance of 70%. The graphene based TCE was demonstrated for organic solar cells. Graphene-based TCE has also been used for the DSSCs [63]. The film thickness of <20 nm, the optical transmittance of >80% and the sheet resistance of the TCE was in the range of 5k to 1M ohm/sq. However, the GO-based method for the preparation of graphene and TCE was having disadvantage due to the high sheet resistance due to the introduction of the functional group on the GO.

CVD grown graphene is having high optical transmittance as well as good sheet resistance which makes the reliable method for the preparation of graphene-based TCEs. In this CVD technique, Cu foil is used as the substrate for the growth of graphene. The carbon solubility is low in Cu foil which limits the precipitation of carbon while cooling down the furnace. The growth of graphene on the Cu foil is mainly due to the catalytic decomposition of hydrocarbon on the surface and subsequent growth of graphene on the Cu substrate. Depending on the conditions, single-crystal or polycrystalline graphene can be grown on the Cu foil. Roll-to-roll production of graphene has been achieved by CVD technique on 30 inch Cu foil [64]. They have doped with p-type doping on graphene. The sheet resistance of ~125 ohm sq^{-1} with T = 97.4% and ~30 ohm sq^{-1} with T = 90%.

5.2 Carbon nanotubes (CNTs) based TCEs

SWCNTs is a one-layer carbon atom thick seamless cylinder bounded together by sp^2 carbon bond [65,66]. Carbon nanotubes are one-dimensional carbon nanomaterials which can be prepared in single-walled (SWCNT) and multi-walled (MWCNT) forms [67]. It is having superior properties such as mobility as high as 100,000 cm^2 V^{-1}s^{-1}, current carrying capacities of 109 A/cm^2 and ON/OFF current ratios as large as 105 [68,69]. SWCNTs also exhibit high surface area to volume ratio and excellent thermal stability. The CNTs can form a network structure as film due to its 1D structure and high aspect ratio [70]. The CNT network films are transparent below 100 nm and having high electrical conductivity. Thus, the CNT network films are widely studied for the possible candidate for the TCE. The CNT-based TCE mainly focused on the SWCNTs because MWCNTs usually accompanies with defect which reduces the conductivity and affects the TCE properties [21-23]. CNTs are produced in high quality using different techniques

Materials for Solar Cell Technologies II   Materials Research Forum LLC
Materials Research Foundations **104** (2021) 77-113   https://doi.org/10.21741/9781644901410-4

such as chemical vapor deposition (CVD), arc-discharge technique and laser ablation technique [71-74]. The formation of SWCNTs and MWCNTs can be controlled by varying the synthesis conditions and parameters. SWCNTs can be formed into network with uniform, optically homogeneous film where thickness can be controlled [75]. The SWCNTs network can be formed by simple filtration technique, air brushing, drop-drying from solvent, Langmuir-Blodgett deposition and lithography techniques [75-80].

For using CNTs as the TCE materials, the impurities such as metal catalysts and carbon impurities need to be thoroughly removed. The purification process of the CNTs consists of removal of the other carbon impurities and defective CNTs by controlled oxidation followed by acid washing to remove the metal impurities which used as catalyst for CNT growth. The harsh purification induces the defects on the surface of the CNTs which leads to the increase in the sheet resistance. Thus, it is important to control the purification procedure for the CNTs to obtain high quality CNT-based TCEs. The purified CNTs need to be deposited on the substrate and mostly solution-based deposition technique is widely used. Due to strong Van der Waals interaction between CNTs, the dispersion of CNTs in solution is realized by surface functionalization on CNTs, by selective dispersion on organic solvents and using surfactants.

SWCNT film has been reported on plastic substrate using line patterned technique with high optical transparency of >80% and low-sheet resistance (<100 Ω/sq) [81]. Similarly, Lee et al. [82] reported the transparent SWCNTs (t-SWCNT) film for the optical transparent p-type ohmic contacts to GaN light-emitting diodes. The t-SWCNT was patterned using standard microlithography technique. A simple process for fabricating SWCNT film by filtration process which was transparent, electrically conductivities has been demonstrated with the transferring of the SWCNT film on various substrates [75].The optical transparency and conductance of the SWCNTs film prepared by the vacuum filtration technique has been found to be dependent on nanotube length. It has been highlighted that the high intertube resistance limits the conductivity properties of the CNT films [83]. Geng and coworkers [84] studied the effect of acid treatment on CNTs on the electrical conductivity without affecting the transmittance. The acid treatment was observed to effectively remove the dispersing agent (sodium dodecyl sulphate) which subsequently found to improve the junction contact by reducing contact resistance.

The CNT-based TCEs are excellent replacement for the ITO due to its flexibility, scalability, solution processability and excellent optical and electrical properties which is very important for the TCEs. However, still, the CNT-based TCEs are limited by the high sheet resistance due to nanotube-nanotube junction contact resistance and roughness. Recently, Jiang et al reported network of isolated SWCNTs by modified CVD technique [85]. The pristine CNT film exhibited the optical transparency of 90% and sheet

Materials for Solar Cell Technologies II   Materials Research Forum LLC
Materials Research Foundations **104** (2021) 77-113   https://doi.org/10.21741/9781644901410-4

resistance of $45\Omega.sq^{-1}$. The acid-treatment using HNO_3 further reduced the sheet resistance to $25\ \Omega.sq^{-1}$ without affecting optical transparency. The observed improvement is due to the welding of SWCNTs by graphitic carbon the intertube junction.

6. Metallic nanowire based TCEs

Metal nanowires have received considerable attention for the development of advanced TCEs because of the inherent electrical conductivity and optical transparency. The main advantages of metal nanowires are the solution-phase synthesis and solution-processable coating of TCE film, which drastically reduces the cost and increases the production capability.

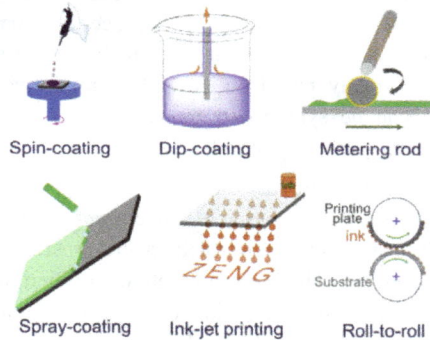

Figure 7. Various processes of high throughput fabrication

6.1 Silver nanowires

Silver nanowires are studied for the TCE applications because of the very low resistivity of the silver in the bulk [86]. Silver nanowire-based TCE was first reported by Lee and co-workers [12] They modelled the optical transmittance of the silver grating in the period of 400 nm, line-width of 40 nm and various heights. They showed that the metal grating is far superior to the ITO TCE with low sheet resistance. By this evidence, they deposited silver nanowire over the glass prepared by polyol synthesis using PVP as the capping agent. The silver nanowires were having the length of 103 ± 17 nm with diameter of 8.7 ± 3.7 nm. The nanowire suspension was drop-casted over the glass. The initial sheet resistance of the as-prepared silver nanowire mesh was $>1\ \Omega.sq^{-1}$ due to the thin-layer of PVP over the silver nanowire. The silver nanowire was heated at 200 °C for 20 nm where the sheet resistance drastically decreased from > 1 to $100\ \Omega.sq^{-1}$ due to the degradation of PVP and silver nanowire-to-silver nanowire contact and fuse together.

Light-induced heating on the junction of the silver nanowires for the plasmonic welding was introduced by Garnett et al. [87]. The small gap between the silver nanowires in the junctions can be used to localized heating by effective light concentration and welding the junction. This reduces the sheet resistance and improves the optical transmittance. The advantage of this technique is self-limited and heat sensitive materials because of the strong dependence of the plasmonic field due to the distance between the wires.

6.2   Copper nanowires

Copper is 1000 times more abundant than silver in the earth's crust and 100 times cheaper than silver and 6000 times cheaper than gold. The conductivity of the copper is slightly lower than the silver (6%). The main challenges in the synthesis of CuNWs are enhanced surface oxidation due to the large surface-to-volume ratio, stability and dispersity. There are various techniques available for the synthesis of copper nanowires. Synthesis of AgNWs is mainly based on the polyol route and it is well-established. However, the CuNWs can be synthesised by various techniques like solution-based technique, electrospinning, hydrothermal [88].

Solution-based synthesis can be broadly classified into aqueous and non-aqueous route for the synthesis of CuNWs. In aqueous route, one approach is the reduction of $CuCl_2$ under pressure at 175°C in water for 10-48 h in the presence of octadecylamine (or oleylamine or hexadecylamine). The low temperature modified method for the synthesis of the above mentioned approach is the reduction of glucose, temperature of 100°C and less than 6 h. This approach is known as alkylamine-mediated synthesis and the solvent used is water. The nanowires with the length >100 nm and diameter around 20 nm which yields high aspect ratio are prepared and is highly desirable for the TCE films. Another approach for the preparation of CuNWs is EDA-mediated synthesis [89]. Chang and coworkers [90] reported the synthesis of CuNWs using EDA with length of 40-50 nm and diameter of 90-120 nm and aspect ratio of >350-450. High concentration of NaOH is necessary to avoid the precipitation of copper hydroxide. Wiley et al. [91] groups have developed EDA-mediated CuNWs are synthesized relatively mild reaction condition at 60°C in alkaline condition. Hydrazine is used as the reducing agent and EDA acts as the structural directing agent. The mechanism of formation of CuNWs is growing from the spherical seed.

Various other methods have been attempted to synthesize the CuNWs, such as non-aqueous route, disproportionation reaction, seeded, imine, amino acid, ascorbic acid, and hydrothermal route. Reduction of their junction resistance has been attempted by post-treatments like thermal annealing, wet chemical coating, nanoplasmonic welding and electric welding. Longer the NWs the junction resistance can be reduced. Cui et al. [92]

have been investigating the development of NWs through electrospinning in various architectures. In order to improve the junction resistance, photonic welding is widely used. CuNW-conducting polymer and transparent polymer composites are developed for the TCEs.

7.   Hybrid nanostructured TCEs

The hybrid nanostructure based TCEs (HNS-TCEs) are featured TCEs that combines and inherit unique properties of their parental constituents at the nanoscale. Therefore, they possess a synergistic effect between individual components, provided they are designed carefully. In the HNS-TCEs the weakness of one component is compensated by the strength of the other component(s). In following sections, such HNS-TCEs are reviewed where physical or chemical limitation of the first component can be taken care by the other.

7.1   Metal nanowires-conductive metal oxide (MNW-MO) hybrids

Metal nanowires (MNWs) are promising for their excellent electrical properties. However, there are some issues which must be resolved before they can be successfully applied to optoelectronic devices. A few of them include i) surface roughness, ii) poor adhesion to substrates, iii) thermal instability, iv) high junction resistance, and v) limited contact area with the adjacent layer [93]. Hybrid structures composed of a metal nanowire mesh and conductive metal-oxide layer have been suggested as promising alternatives to resolve these issues [94]. It has been seen that incorporation of the conductive metal oxide into the host MNWs protects the nanowires from local melting-induced disconnection, which in turn enhances the thermal stability of the MNWs. In addition to this the metal-oxide overcoating layers improve the adhesion property of the MNWs to the substrate. Such composite electrode also facilitates effective charge transfer by filling the empty space unoccupied by the MNWs and by smoothing the electrode surfaces [95]. Table 4 [93, 96-98] lists some notable developments in literature as on date of MNW-MO based TCEs.

7.2   Metal nanowires-graphene (MNW-graphene) hybrids

As mentioned earlier, MNW networks have certain limitations towards commercialization. They exhibit high NW-NW contact resistance, high surface roughness, high oxidation probability (especially in CuNW) and poor adhesion to plastic substrates. The CVD graphene films are usually polycrystalline with grain boundaries that affect the charge mobility and hence the electrical conductivity. The transferred graphene electrodes also have intrinsic defects and dislocations leading to their conductivity lower than that of exfoliated graphene. However, combining graphene and MNWs overcomes the weakness of individual component. In some cases, MNW-core/graphene or reduced graphene oxide (RGO)-shell structure has been shown to protect the MNW such as CuNW from oxidation [97]. Table 4 [99-101] lists some important developments in the literature as on date for MNW-graphene/GO based TCEs.

Table 4. TCE parameters reported in some hybrid nanostructured systems

HNS-TCE system	HNS-TCE example	Sheet resistance (Rs) $\Omega.sq^{-1}$	Transmittance (%) at 550 nm	Research group/ year/ Reference
MNW/MO	ITO/AgNW	23.0	88.6	Yang/ 2012/ [96]
	AZO/AgNW/AZO-ZnO	11.3	93.4	Moon/ 2014/ [93]
	IZO/ AgNW	9.3	91.0	Subramanian/ 2016/ [97]
	SnO$_x$/AgNW	5.2	81.0	Riedel/ 2014/ [98]
MNW-Graphene	CVD graphene/AgNW	16	91.1	Zhang group/ 2014/ [99]
	spin coated AgNWs/CVD-graphene	33	94.0	Park group/ 2013/ [100]
	CuNW/RGO core-shell	28	90	Yang group/ 2016/ [101]
MNW-CNT	AgNW/SWNT (3:2)	120	90.8	Han group/ 2015/ [102]
Graphene-CNT	(CVD) Graphene/ (CVD) SWNT	420	~ 73	Tai group/ 2012/ [103]
	RGO/ SWNT	180	86	Kim group/ 2012/ [104]
Polymer composite based (CPC)	PEDOT:PSS/Graphene	0.2 S / cm (Conductivity)	~ 80	Chen group/ 2009/ [105] Yang group/ 2011/ [106]

Abbreviations: AZO – Al doped ZnO, ITO – In doped SnO2, IZO – In doped ZnO, RGO – Reduced Graphene Oxide

7.3   Metal nanowires-carbon nanotube (MNW-CNT) hybrids

The main advantage of integrating MNW with SWCNTs has been the overcoming of the high haze factor in some MNWs, such as AgNW [102]. By the plasma resonance property of SWCNTs, they absorb light scattered from the AgNWs, which in turn reduces the haze of such MNWs. However, such composites have not gained much attention due to tricky optimization and trade-off between the sheet resistance and the haze values.

7.4   Graphene-carbon nanotube (graphene-CNT) hybrids

Although graphene is a proven material with outstanding electrical mobility and very high transmittance to become the material of choice for TCEs, the practical issue associated with graphene deposition on the substrate is the domain size limitations and their interconnectivity. This leads to worsening its electrical merit. Carbon nanotubes can synergistically solve this problem to act as domain- to -domain interconnecting media. Typically, graphene sheet resistance has been shown to improve from 2.15 to 420 $\Omega.sq^{-1}$ by CVD grown SWNTs. Although, the transmittance in such cases has been found to get affected from 85 to 73%. Therefore, density of spatial coverage of CNTs plays a crucial role on determining the transparency. Moreover, SWNTs have also been used with reduced graphene oxides (RGO), where introducing them between the RGO sheets is a sustainable approach for inhibiting the agglomeration of sheets after chemical reduction takes place. Table 4 [103,104] lists some of the best reported TCEs based on graphene-CNT composites.

7.5   Conductive polymer (CP) composite transparent electrodes

A conductive polymer matrix, such as PEDOT:PSS may be used as a bridging medium in the nanosheet-to-nanosheet electrical conduction in the graphene films such as by the CNTs as mentioned in the foregoing section. In addition to that, a CP can also act as flexible host and therefore, such CP-graphene composite can act as promising CP-composite TCE [105]. The CP has also been useful in MNW based TCEs. With organic polymers such as, PEDOT:PSS or amyl acetate (nail polish solvent) MNWs such as AgNWs have shown excellent adhesion property with the glass as well as flexible substrates [106]. Table 4 [105,106] enlists some reported TCE characteristics of such CP composites.

7.6   Graphene-metal oxide (graphene-MO) hybrids

As mentioned above, graphene needs a bridging matrix to reduce its sheet resistance originated from dispersed nanosheets over the substrate, amid exceptionally high electronic mobility. Some metal oxide (MO), such as very thin layer of silica can serve as protective host as well [107].

8   Scalability and large area fabrication of advanced TCEs

Large-area fabrication of the advanced TCEs is important for the commercial applications. It is well-known that the solution-based deposition technique is considered promising for the large-area fabrication due to its low-cost, scalability and high-throughput. In solution-based deposition technique, the TCE materials are uniformly dispersed in the suitable solvents and the dispersed solution is subsequently coated onto the substrate. Some of the solution-based deposition techniques are vacuum filtration, spin-coating, dip-coating, spray coating, rod coating and ink-jet print coating.

The vacuum filtration method is a solution-based deposition technique for the thin film production. In this deposition technique, vacuum filtration is used for the deposition of thin film on the filtration membrane followed by transferring the deposited film on the targeted substrate from the filter membrane. The TCEs produced from this method exhibit highly homogenous and the process parameters can be used to control the properties of the TCEs by simply varying the concentration and volume of the solution. By using filtration techniques, CNTs, rGO, AgNWs and CuNWs and their hybrids are widely reported. The main limitation of the vacuum filtration technique is its scalability. This method is restricted to the development of small area TCE films. Similarly, another solution-based technique is the spin coating for fabricating uniform TCEs on the flat substrates. The properties of the TCEs are controlled by the rotation speed, duration, concentration of the solution, etc. However, this method is also limited for the scalability and large-area deposition.

Some of the scalable methods for the fabrication of large-area TCEs are spray coating, rod coating (Meyer rod) and ink-jet printing technique. These techniques can be adopted for the continuous production of TCEs by roll-by-roll coating. These are all solution-based deposition techniques which are simple, fast, low-cost, and scalable deposition method to obtain TCEs. The method can be adopted for the room temperature and different substrates can be used. In the spray coating method, the TC materials can be directly coated on the substrate by spraying the solution using air gun or ultrasonic spray gun. The advantage of this method is deposition of materials on the different types of substrate and in different geometries. The Meyer rod coating is a solution-based deposition method which is scalable and low-cost technique. This is a continuous roll-by-roll fabrication TCE fabrication technique in which high quality TCEs can be deposited on the substrate. The properties of the film can be controlled by the viscosity of the coating slurry, speed of coating and wetting property. Mainly, metal nanowire-based TCE fabrication is widely reported by this technique. The ink-jet printing method is considered as the fast and low-cost printing technique because this method does not have any templates or mask for fabrication of thin TCE films. This method can be adopted for the large-area and roll-to-roll production of films.

9. Prospects

ITO is widely used TCEs in the commercial market for almost six decades and it is expected that the ITO will be continued to be used in the future. However, rapid growth in flexible, bendable, and stretchable devices in coming decades necessitates the development of the advanced TCEs. Also, availability of the indium and cost of manufacturing the oxide-based TCEs by vapor-based technique are driving to find suitable replacement of ITO.

CNT-based TCEs have made huge progress with the optical transparency and electrical conductivity comparable to the ITO. CNT film exhibits excellent conductivity, mechanical flexibility and required optical properties. However, the TCEs-based on CNT still need to reduce the sheet resistance which is high due to the contact resistance between the nanotubes and high roughness factor. Graphene is another promising candidate for the TCE. The SLG is having the optical transparency of ~97%. However, obtaining the continuous single-layer graphene without defects is limiting the large-scale fabrication and continuous roll-by-roll production. In this aspect, metal nanowires (NWs) such as AgNWs and CuNWs are most promising candidates which can be produced in large scale and processed at room temperature. The large-scale fabrication is possible by using Meyer rod coating, spray coating and ink-jet printing methods which enables to produce continuous roll-by-roll coating of metal nanowire-based TCEs. This also enables the fabrication of TCEs at lower-cost when compared to vapor-based fabrication of ITOs. AgNW-based TCEs are already available in the commercial market. However, the cost of Ag limits the future of the AgNW-based TCE. In this aspect, CuNW-based TCEs is having potential for the future TCEs because of the low-cost, scalable and good mechanical, electrical and optical properties. However, the CuNWs are prone to oxidation in the ambient condition. Recently, hybrid TCEs are widely studied for the TCEs and these hybrid TCEs need to be fabricated in the large-scale.

References

[1]   H.C. Chu, Y.C. Chang, Y. Lin, S.H. Chang, W.C. Chang, G.A. Li, H.Y. Tuan, Spray-deposited large-area copper nanowire transparent conductive electrodes and

their uses for touch screen applications, ACS Appl. Mater. Interfaces 8 (2016) 13009-13017. doi: 10.1021/acsami.6b02652

[2]   A.R. Madaria, A. Kumar, C. Zhou, Large scale, highly conductive and patterned transparent films of silver nanowires on arbitrary substrates and their application in touch screens, Nanotechnology 22 (2011) 245201. doi: 10.1088/0957-4484/22/24/245201

[3]   T. Minami, S. Takata, T. Kakumu, New multicomponent transparent conducting oxide films for transparent electrodes of flat panel displays, J. Vac. Sci. Technol. A 14 (1996) 1689-1693. doi: 10.1116/1.580320

[4]   G.S. Chae, A modified transparent conducting oxide for flat panel displays only, Jpn. J. Appl. Phys. 40 (2001) 1282. doi: 10.1143/JJAP.40.1282

[5]   G. Jo, M. Choe, C.Y. Cho, J.H. Kim, W. Park, S. Lee, W.-K. Hong, T.-W. Kim, S.-J. Park, B.H. Hong, Large-scale patterned multi-layer graphene films as transparent conducting electrodes for GaN light-emitting diodes, Nanotechnology 21 (2010) 175201. doi: 10.1088/0957-4484/21/17/175201

[6]   H. Liu, V. Avrutin, N. Izyumskaya, Ü. Özgür, H. Morkoç, Transparent conducting oxides for electrode applications in light emitting and absorbing devices, Superlattices Microstruct. 48 (2010) 458-484. doi: 10.1016/j.spmi.2010.08.011

[7]   K. Rana, J. Singh, J.H. Ahn, A graphene-based transparent electrode for use in flexible optoelectronic devices, J. Mater. Chem. C 2 (2014) 2646-2656. doi: 10.1039/C3TC32264E

[8]   D. Langley, G. Giusti, C. Mayousse, C. Celle, D. Bellet, J.P. Simonato, Flexible transparent conductive materials based on silver nanowire networks: a review, Nanotechnology 24 (2013) 452001. doi: 10.1088/0957-4484/24/45/452001

[9]   A.W. Wright, ART. VII.--on the production of transparent metallic films by the electrical discharge in exhausted tubes, American Journal of Science and Arts (1820-1879) 13 (1877) 49.

[10]   K. Baedeker, Über die elektrische Leitfähigkeit und die thermoelektrische Kraft einiger Schwermetallverbindungen, Annalen der Physik 327 (1907): 749-766. doi:10.1002/andp.19073270409

[11]   T. Minami, Transparent conducting oxide semiconductors for transparent electrodes, Semicond. Sci. Technol. 20 (2005) S35. doi: 10.1088/0268-1242/20/4/004

[12]   H.B. Lee, W.-Y. Jin, M.M. Ovhal, N. Kumar, J.W. Kang, Flexible transparent conducting electrodes based on metal meshes for organic optoelectronic device

applications: a review, J. Mater. Chem. C 7 (2019) 1087-1110. doi:
10.1039/C8TC04423F

[13]   J. Lewis, S. Grego, B. Chalamala, E. Vick, D. Temple, Highly flexible transparent
electrodes for organic light-emitting diode-based displays, Appl. Phys. Lett. 85 (2004)
3450-3452. doi: 10.1063/1.1806559

[14]   K.H. Kim, B.R. Koo, H.J. Ahn, Sheet resistance dependence of fluorine-doped tin
oxide films for high-performance electrochromic devices, Ceram. Int. 44 (2018) 9408-
9413. doi: 10.1016/j.ceramint.2018.02.157

[15]   R. Shukla, A. Srivastava, A. Srivastava, K. Dubey, Growth of transparent
conducting nanocrystalline Al doped ZnO thin films by pulsed laser deposition, J.
Cryst. Growth 294 (2006) 427-431. doi: 10.1016/j.jcrysgro.2006.06.035

[16]   T. Sannicolo, M. Lagrange, A. Cabos, C. Celle, J.P. Simonato, D. Bellet, Metallic
nanowire-based transparent electrodes for next generation flexible devices: a Review,
Small 12 (2016) 6052-6075. doi: 10.1002/smll.201602581

[17]   M. Vosgueritchian, D.J. Lipomi, Z. Bao, Highly conductive and transparent
PEDOT: PSS films with a fluorosurfactant for stretchable and flexible transparent
electrodes, Adv. Funct. Mater. 22 (2012) 421-428. doi: 10.1002/adfm.201101775

[18]   J. Wu, M. Agrawal, H.A. Becerril, Z. Bao, Z. Liu, Y. Chen, P. Peumans, Organic
light-emitting diodes on solution-processed graphene transparent electrodes, ACS
Nano 4 (2010) 43-48. doi: 10.1021/nn900728d

[19]   Y. Xu, J. Liu, Graphene as transparent electrodes: fabrication and new emerging
applications, Small 12 (2016) 1400-1419. doi: 10.1002/smll.201502988

[20]   Y. Lee, J.H. Ahn, Graphene-based transparent conductive films, Nano 8 (2013)
1330001. doi: 10.1142/S1793292013300016

[21]   S. Park, M. Vosguerichian, Z. Bao, A review of fabrication and applications of
carbon nanotube film-based flexible electronics, Nanoscale 5 (2013) 1727-1752. doi:
10.1039/C3NR33560G

[22]   A.J. Stapleton, R.A. Afre, A.V. Ellis, J.G. Shapter, G.G. Andersson, J.S. Quinton,
D.A. Lewis, Highly conductive interwoven carbon nanotube and silver nanowire
transparent electrodes, Sci. Technol. Adv. Mater. 14 (2013) 035004. doi:
10.1088/1468-6996/14/3/035004

[23]   Y.M. Chien, F. Lefevre, I. Shih, R. Izquierdo, A solution processed top emission
OLED with transparent carbon nanotube electrodes, Nanotechnology 21 (2010)
134020. doi: 10.1088/0957-4484/21/13/134020

[24]   N. Saran, K. Parikh, D.-S. Suh, E. Munoz, H. Kolla, S.K. Manohar, Fabrication and characterization of thin films of single-walled carbon nanotube bundles on flexible plastic substrates, J. Am. Chem. Soc. 126 (2004) 4462-4463. doi: 10.1021/ja037273p

[25]   B. Deng, P.C. Hsu, G. Chen, B. Chandrashekar, L. Liao, Z. Ayitimuda, J. Wu, Y. Guo, L. Lin, Y. Zhou, Roll-to-roll encapsulation of metal nanowires between graphene and plastic substrate for high-performance flexible transparent electrodes, Nano Lett. 15 (2015) 4206-4213. doi: 10.1021/acs.nanolett.5b01531

[26]   K. Zilberberg, T. Riedl, Metal-nanostructures–a modern and powerful platform to create transparent electrodes for thin-film photovoltaics, J. Mater. Chem. A 4 (2016) 14481-14508. doi: 10.1039/C6TA05286J

[27]   D. Zhang, R. Wang, M. Wen, D. Weng, X. Cui, J. Sun, H. Li, Y. Lu, Synthesis of ultralong copper nanowires for high-performance transparent electrodes, J. Am. Chem. Soc. 134 (2012) 14283-14286. doi:10.1021/ja3050184

[28]   S. Ye, A.R. Rathmell, Z. Chen, I.E. Stewart, B.J. Wiley, Metal nanowire networks: the next generation of transparent conductors, Adv. Mater. 26 (2014) 6670-6687. doi: 10.1002/adma.201402710

[29]   K. Ellmer, Past achievements and future challenges in the development of optically transparent electrodes, Nature Photonics 6 (2012) 809. doi: 10.1038/nphoton.2012.282

[30]   D.S. Hecht, L. Hu, G. Irvin, Emerging transparent electrodes based on thin films of carbon nanotubes, graphene, and metallic nanostructures, Adv. Mater. 23 (2011) 1482-1513. doi: 10.1002/adma.201003188

[31]   S. Kang, T. Kim, S. Cho, Y. Lee, A. Choe, B. Walker, S.J. Ko, J.Y. Kim, H. Ko, Capillary printing of highly aligned silver nanowire transparent electrodes for high-performance optoelectronic devices, Nano Lett. 15 (2015) 7933-7942. doi: 10.1021/acs.nanolett.5b03019

[32]   Z. Yu, L. Li, Q. Zhang, W. Hu, Q. Pei, Silver nanowire-polymer composite electrodes for efficient polymer solar cells, Adv. Mater. 23 (2011) 4453-4457. doi: 10.1002/adma.201101992

[33]   X.Y. Zeng, Q.K. Zhang, R.M. Yu, C.Z. Lu, A new transparent conductor: silver nanowire film buried at the surface of a transparent polymer, Adv. Mater. 22 (2010) 4484-4488. doi: 10.1002/adma.201001811

[34]   D.S. Leem, A. Edwards, M. Faist, J. Nelson, D.D. Bradley, J.C. De Mello, Efficient organic solar cells with solution-processed silver nanowire electrodes, Adv. Mater. 23 (2011) 4371-4375. doi: 10.1002/adma.201100871

[35]   S. Yun, X. Niu, Z. Yu, W. Hu, P. Brochu, Q. Pei, Compliant silver nanowire-polymer composite electrodes for bistable large strain actuation, Adv. Mater. 24 (2012) 1321-1327. doi: 10.1002/adma.201104101

[36]   H.G. Im, S.H. Jung, J. Jin, D. Lee, J. Lee, D. Lee, J.Y. Lee, I.D. Kim, B.S. Bae, Flexible transparent conducting hybrid film using a surface-embedded copper nanowire network: a highly oxidation-resistant copper nanowire electrode for flexible optoelectronics, ACS Nano 8 (2014) 10973-10979. doi: 10.1021/nn504883m

[37]   V.B. Nam, D. Lee, Copper nanowires and their applications for flexible, transparent conducting films: a review, Nanomaterials 6 (2016) 47. doi: 10.3390/nano6030047

[38]   C. Sachse, N. Weiß, N. Gaponik, L. M. Meskamp, A. Eychmüller, K. Leo, ITO - free, small-molecule organic solar cells on spray-coated copper-nanowire-based transparent electrodes, Adv. Energy Mater. 4 (2014) 1300737. doi: 10.1002/aenm.201300737

[39]   A.R. Rathmell, S.M. Bergin, Y.L. Hua, Z.Y. Li, B.J. Wiley, The growth mechanism of copper nanowires and their properties in flexible, transparent conducting films, Adv. Mater. 22 (2010) 3558-3563. doi: 10.1002/adma.201000775

[40]   K. Ghaffarzadeh, R. Das, Transparent Conductive Films and Materials 2018-2028: Forecasts, Technologies, Players, IDTechEx Research, 2018.

[41]   G. Hautier, A. Miglio, G. Ceder, G.M. Rignanese, X. Gonze, Identification and design principles of low hole effective mass p-type transparent conducting oxides, Nat. Commun. 4 (2013) 1-7. doi: 10.1038/ncomms3292

[42]   K. Alberi, M.B. Nardelli, A. Zakutayev, L. Mitas, S. Curtarolo, A. Jain, M. Fornari, N. Marzari, I. Takeuchi, M.L. Green, The 2019 materials by design roadmap, J. Phys. D: Appl. Phys. 52 (2018) 013001. doi: 10.1088/1361-6463/aad926

[43]   L. Hu, H.S. Kim, J.Y. Lee, P. Peumans, Y. Cui, Scalable coating and properties of transparent, flexible, silver nanowire electrodes, ACS Nano 4 (2010) 2955-2963. doi: 10.1021/nn1005232

[44]   K. Azuma, K. Sakajiri, H. Matsumoto, S. Kang, J. Watanabe, M. Tokita, Facile fabrication of transparent and conductive nanowire networks by wet chemical etching

with an electrospun nanofiber mask template, Mater. Lett. 115 (2014) 187-189. doi: 10.1016/j.matlet.2013.10.054

[45]   F. Cicoira, C. Santato, Organic electronics: emerging concepts and technologies, John Wiley & Sons (2013). ISBN (Print): 978-3-527-41131-3.

[46]   K.S. Novoselov, V. Fal, L. Colombo, P. Gellert, M. Schwab, K. Kim, A roadmap for graphene, Nature 490 (2012) 192-200. doi: 10.1038/nature11458.

[47]   Please refer to https://www.alliedmarketresearch.com/transparent-conductive-films-market

[48]   C. Kittel, P. McEuen, P. McEuen, Introduction to solid state physics, Wiley New York (1996). ISBN (Print): 978-81-265-1045-0.

[49]   P.R. Wallace, The band theory of graphite, Phys. Rev. 71 (1947) 622. doi: 10.1103/PhysRev.71.622

[50]   Partially adapted from https://physicscatalyst.com/chemistry/modern-periodic-table.png

[51]   K. Shimakawa, S. Narushima, H. Hosono, H. Kawazoe, Electronic transport in degenerate amorphous oxide semiconductors, Philos. Mag. Lett. 79 (1999) 755-761. doi: 10.1080/095008399176823

[52]   Ü. Özgür, Y.I. Alivov, C. Liu, A. Teke, M. Reshchikov, S. Doğan, V. Avrutin, S.J. Cho, Morkoç, A comprehensive review of ZnO materials and devices, J. Appl. Phys. 98 (2005) 11. doi: 10.1063/1.1992666

[53]   H. Hosono, N. Kikuchi, N. Ueda, H. Kawazoe, K.i. Shimidzu, Amorphous transparent electroconductor 2CdO· GeO$_2$: Conversion of amorphous insulating cadmium germanate by ion implantation, Appl. Phys. Lett. 67 (1995) 2663-2665. doi: 10.1063/1.114329

[54]   S. Narushima, M. Orita, M. Hirano, H. Hosono, Electronic structure and transport properties in the transparent amorphous oxide semiconductor 2 CdO· GeO$_2$, Phys. Rev. B 66 (2002) 035203. doi: 10.1103/PhysRevB.66.035203

[55]   K. Nomura, H. Ohta, A. Takagi, T. Kamiya, M. Hirano, H. Hosono, Room-temperature fabrication of transparent flexible thin-film transistors using amorphous oxide semiconductors, Nature 432 (2004) 488-492. doi: 10.1038/nature03090

[56]   H. Hosono, Ionic amorphous oxide semiconductors: Material design, carrier transport, and device application, J. Non-Cryst. Solids 352 (2006) 851-858. doi: 10.1016/j.jnoncrysol.2006.01.073

Materials Research Forum LLC
https://doi.org/10.21741/9781644901410-4

[57] M.P. Taylor, D.W. Readey, M.F. van Hest, C.W. Teplin, J.L. Alleman, M.S. Dabney, L.M. Gedvilas, B.M. Keyes, B. To, J.D. Perkins, The remarkable thermal stability of amorphous In-ZnO transparent conductors, Adv. Funct. Mater. 18 (2008) 3169-3178. doi: 10.1002/adfm.200700604

[58] Y. Furubayashi, T. Hitosugi, Y. Yamamoto, K. Inaba, G. Kinoda, Y. Hirose, T. Shimada, T. Hasegawa, A transparent metal: Nb-doped anatase TiO_2, Appl. Phys. Lett. 86 (2005) 252101. doi: 10.1063/1.1949728

[59] S. Sheng, G. Fang, C. Li, S. Xu, X. Zhao, p-type transparent conducting oxides, physica status solidi (a) 203 (2006) 1891-1900. doi: 10.1002/pssa.200521479

[60] H. Kawazoe, M. Yasukawa, M. Hyodo, M. Kurita, H. Yanagi, H. Hosono, P-type electrical conduction in transparent thin films of $CuAlO_2$, Nature 389 (1997) 939-942. doi: 10.1038/40087

[61] K.S. Kim, Y. Zhao, H. Jang, S.Y. Lee, J.M. Kim, K.S. Kim, J.H. Ahn, P. Kim, J.Y. Choi, B.H. Hong, Large-scale pattern growth of graphene films for stretchable transparent electrodes, Nature 457 (2009) 706-710. doi: 10.1038/nature07719

[62] J. Wang, M. Liang, Y. Fang, T. Qiu, J. Zhang, L. Zhi, Rod-coating: towards large-area fabrication of uniform reduced graphene oxide films for flexible touch screens, Adv. Mater. 24 (2012) 2874-2878. doi: 10.1002/adma.201200055

[63] X. Wang, L. Zhi, K. Müllen, Transparent, conductive graphene electrodes for dye-sensitized solar cells, Nano Lett. 8 (2008) 323-327. doi: 10.1021/nl072838r

[64] S. Bae, H. Kim, Y. Lee, X. Xu, J.S. Park, Y. Zheng, J. Balakrishnan, T. Lei, H.R. Kim, Y.I. Song, Roll-to-roll production of 30-inch graphene films for transparent electrodes, Nat. Nanotechnol. 5 (2010) 574. doi: 10.1038/nnano.2010.132

[65] S. Iijima, Helical microtubules of graphitic carbon, Nature 354 (1991) 56-58. doi: 10.1038/354056a0

[66] H. Zhu, C. Xu, D. Wu, B. Wei, R. Vajtai, P. Ajayan, Direct synthesis of long single-walled carbon nanotube strands, Science 296 (2002) 884-886. doi: 10.1126/science.1066996

[67] M.S. Dresselhaus, G. Dresselhaus, P. Eklund, A. Rao, Carbon nanotubes. In: W. Andreoni (eds) The physics of fullerene-based and fullerene-related materials. physics and chemistry of materials with low-dimensional structures, 23 (2000) 331-379. Springer, Dordrecht. doi: 10.1007/978-94-011-4038-6_9

[68]   W. De Heer, J.M. Bonard, T. Stöckli, A. Chatelain, L. Forro, D. Ugarte, Carbon nanotubes films: electronic properties and their application as field emitters, Z. Phys. D Atom. Mol. Cl. 40 (1997) 418-420. doi: 10.1007/s004600050241

[69]   T.W. Odom, J.-L. Huang, P. Kim, C.M. Lieber, Atomic structure and electronic properties of single-walled carbon nanotubes, Nature 391 (1998) 62-64. doi: 10.1038/34145

[70]   E. Snow, J. Novak, P. Campbell, D. Park, Random networks of carbon nanotubes as an electronic material, Appl. Phys. Lett. 82 (2003) 2145-2147. doi: 10.1063/1.1564291

[71]   H. Dai, Carbon nanotubes: synthesis, integration, and properties, Accounts Chem. Res. 35 (2002) 1035-1044. doi: 10.1021/ar0101640

[72]   J. Prasek, J. Drbohlavova, J. Chomoucka, J. Hubalek, O. Jasek, V. Adam, R. Kizek, Methods for carbon nanotubes synthesis, J. Mater. Chem. 21 (2011) 15872-15884. doi: 10.1039/C1JM12254A

[73]   R. Andrews, D. Jacques, D. Qian, T. Rantell, Multiwall carbon nanotubes: synthesis and application, Accounts Chem. Res. 35 (2002) 1008-1017. doi: 10.1021/ar010151m

[74]   S.B. Sinnott, R. Andrews, Carbon nanotubes: synthesis, properties, and applications, Crit. Rev. Solid State 26 (2001) 145-249. doi: 10.1080/20014091104189

[75]   A. Kaskela, A.G. Nasibulin, M.Y. Timmermans, B. Aitchison, A. Papadimitratos, Y. Tian, Z. Zhu, H. Jiang, D.P. Brown, A. Zakhidov, Aerosol-synthesized SWCNT networks with tunable conductivity and transparency by a dry transfer technique, Nano Lett. 10 (2010) 4349-4355. doi: 10.1021/nl101680s

[76]   C.S. Woo, C.H. Lim, C.W. Cho, B. Park, H. Ju, D.H. Min, C.J. Lee, S.B. Lee, Fabrication of flexible and transparent single-wall carbon nanotube gas sensors by vacuum filtration and poly (dimethyl siloxane) mold transfer, Microelectronic. Eng. 84 (2007) 1610-1613. doi: 10.1016/j.mee.2007.01.162

[77]   M. Kaempgen, G. Duesberg, S. Roth, Transparent carbon nanotube coatings, Appl. Surf. Sci. 252 (2005) 425-429. doi: 10.1016/j.apsusc.2005.01.020

[78]   F. Antolini, T. Di Luccio, E. Serra, P. Aversa, L. Tapfer, S. Sangiorgi, Deposition and characterization of Langmuir-Blodgett films of cadmium arachidate/SWCNTs composites, Surf. Interface Anal. 38 (2006) 1285-1290. doi: 10.1002/sia.2391

[79]   Y.L. Tai, Z.G. Yang, Flexible, transparent, thickness-controllable SWCNT/PEDOT: PSS hybrid films based on coffee-ring lithography for functional

noncontact sensing device, Langmuir 31 (2015) 13257-13264. doi:
10.1021/acs.langmuir.5b03449

[80]   S. Ravi, A.B. Kaiser, C.W. Bumby, Improved conduction in transparent single walled carbon nanotube networks drop-cast from volatile amine dispersions, Chem. Phy. Lett. 496 (2010) 80-85. doi: 10.1016/j.cplett.2010.06.084

[81]   G. Gruner, Carbon nanotube films for transparent and plastic electronics, J. Mater. Chem. 16 (2006) 3533-3539. doi: 10.1039/B603821M

[82]   K. Lee, Z. Wu, Z. Chen, F. Ren, S. Pearton, A. Rinzler, Single wall carbon nanotubes for p-type ohmic contacts to GaN light-emitting diodes, Nano Letters 4 (2004) 911-914. doi:

[83]   D. Simien, J.A. Fagan, W. Luo, J.F. Douglas, K. Migler, J. Obrzut, Influence of nanotube length on the optical and conductivity properties of thin single-wall carbon nanotube networks, ACS Nano 2 (2008) 1879-1884. doi: 10.1021/nn800376x

[84]   H.Z. Geng, K.K. Kim, Y.H. Lee, Recent progress in carbon nanotube-based flexible transparent conducting film, Carbon nanotubes and associated devices, carbon nanotubes and associated devices, Int. Soc. Optics and Photonics 7037 (2008) 70370A. doi: 10.1117/12.796143

[85]   C. Jiang, J. Zhao, H.A. Therese, M. Friedrich, A. Mews, Raman imaging and spectroscopy of heterogeneous individual carbon nanotubes, J. Phys. Chem. B 107 (2003) 8742-8745. doi: 10.1021/jp035371r

[86]   S. De, T.M. Higgins, P.E. Lyons, E.M. Doherty, P.N. Nirmalraj, W.J. Blau, J.J. Boland, J.N. Coleman, Silver nanowire networks as flexible, transparent, conducting films: extremely high DC to optical conductivity ratios, ACS Nano 3 (2009) 1767-1774. doi: 10.1021/nn900348c

[87]   E.C. Garnett, W. Cai, J.J. Cha, F. Mahmood, S.T. Connor, M.G. Christoforo, Y. Cui, M.D. McGehee, M.L. Brongersma, Self-limited plasmonic welding of silver nanowire junctions, Nat. Mater. 11 (2012) 241-249. doi: 10.1038/nmat3238

[88]   A.R. Rathmell, B.J. Wiley, The synthesis and coating of long, thin copper nanowires to make flexible, transparent conducting films on plastic substrates, Adv. Mater. 23 (2011) 4798-4803. doi: 10.1002/adma.201102284

[89]   Y.S. Cho, Y.D. Huh, Synthesis of ultralong copper nanowires by reduction of copper-amine complexes, Mater. Lett. 63 (2009) 227-229. doi: 10.1016/j.matlet.2008.09.049

[90]   Y. Chang, M.L. Lye, H.C. Zeng, Large-scale synthesis of high-quality ultralong copper nanowires, Langmuir 21 (2005) 3746-3748. doi: 10.1021/la050220w

[91]   S. Ye, A.R. Rathmell, I.E. Stewart, Y.-C. Ha, A.R. Wilson, Z. Chen, B.J. Wiley, A rapid synthesis of high aspect ratio copper nanowires for high-performance transparent conducting films, Chem. Commun. 50 (2014) 2562-2564. doi: 10.1039/C3CC48561G

[92]   P.C. Hsu, D. Kong, S. Wang, H. Wang, A.J. Welch, H. Wu, Y. Cui, Electrolessly deposited electrospun metal nanowire transparent electrodes, J. Am. Chem. Soc. 136 (2014) 10593-10596. doi: 10.1021/ja505741e

[93]   A. Kim, Y. Won, K. Woo, S. Jeong, J. Moon, All-solution-processed indium-free transparent composite electrodes based on Ag nanowire and metal oxide for thin-film solar cells, Adv. Func. Mater. 24 (2014) 2462-2471. doi: 10.1002/adfm.201303518

[94]   A. Kim, Y. Won, K. Woo, C.-H. Kim, J. Moon, Highly transparent low resistance ZnO/Ag nanowire/ZnO composite electrode for thin film solar cells, ACS Nano 7 (2013) 1081-1091. doi: 10.1021/nn305491x

[95]   T. Stubhan, J. Krantz, N. Li, F. Guo, I. Litzov, M. Steidl, M. Richter, G.J. Matt, C.J. Brabec, High fill factor polymer solar cells comprising a transparent, low temperature solution processed doped metal oxide/metal nanowire composite electrode, Sol. Energy Mater. Sol. Cells 107 (2012) 248-251. doi: 10.1016/j.solmat.2012.06.039

[96]   C.H. Chung, T.B. Song, B. Bob, R. Zhu, Y. Yang, Solution-processed flexible transparent conductors composed of silver nanowire networks embedded in indium tin oxide nanoparticle matrices, Nano Res. 5 (2012) 805-814. doi: 10.1007/s12274-012-0264-8

[97]   W.J. Scheideler, J. Smith, I. Deckman, S. Chung, A.C. Arias, V. Subramanian, A robust, gravure-printed, silver nanowire/metal oxide hybrid electrode for high-throughput patterned transparent conductors, J. Mater. Chem. C 4 (2016) 3248-3255. doi: 10.1039/C5TC04364F

[98]   K. Zilberberg, F. Gasse, R. Pagui, A. Polywka, A. Behrendt, S. Trost, R. Heiderhoff, P. Görrn, T. Riedl, Highly robust indium-free transparent conductive electrodes based on composites of silver nanowires and conductive metal oxides, Adv. Func. Mater. 24 (2014) 1671-1678. doi: 10.1002/adfm.201303108

[99]   S. Xu, B. Man, S. Jiang, M. Liu, C. Yang, C. Chen, C. Zhang, Graphene–silver nanowire hybrid films as electrodes for transparent and flexible loudspeakers, Cryst. Eng. Comm. 16 (2014) 3532-3539. doi: 10.1039/c3ce42656d

[100] M.S. Lee, K. Lee, S.-Y. Kim, H. Lee, J. Park, K.-H. Choi, H.-K. Kim, D.G. Kim, D.Y. Lee, S. Nam, High-performance, transparent, and stretchable electrodes using graphene–metal nanowire hybrid structures, Nano Lett. 13 (2013) 2814-2821. doi: 10.1021/nl401070p

[101] L. Dou, F. Cui, Y. Yu, G. Khanarian, S.W. Eaton, Q. Yang, J. Resasco, C. Schildknecht, K. Schierle-Arndt, P. Yang, Solution-processed copper/reduced-graphene-oxide core/shell nanowire transparent conductors, ACS Nano 10 (2016) 2600-2606. doi: 10.1021/acsnano.5b07651

[102] H.J. Han, Y.C. Choi, J.H. Han, Preparation of transparent conducting films with improved haze characteristics using single-wall carbon nanotube-silver nanowire hybrid material, Synthetic Met. 199 (2015) 219-222. doi: 10.1016/j.synthmet.2014.11.014

[103] D.D. Nguyen, N.-H. Tai, S.Y. Chen, Y.L. Chueh, Controlled growth of carbon nanotube–graphene hybrid materials for flexible and transparent conductors and electron field emitters, Nanoscale 4 (2012) 632-638. doi: 10.1039/C1NR11328C

[104] Z.D. Huang, B. Zhang, S.W. Oh, Q.B. Zheng, X.Y. Lin, N. Yousefi, J.K. Kim, Self-assembled reduced graphene oxide/carbon nanotube thin films as electrodes for supercapacitors, J. Mater. Chem. 22 (2012) 3591-3599. doi: 10.1039/C2JM15048D

[105] Y. Xu, Y. Wang, J. Liang, Y. Huang, Y. Ma, X. Wan, Y. Chen, A hybrid material of graphene and poly (3, 4-ethyldioxythiophene) with high conductivity, flexibility, and transparency, Nano Res. 2 (2009) 343-348. doi: 10.1007/s12274-009-9032-9

[106] R. Zhu, C.-H. Chung, K.C. Cha, W. Yang, Y.B. Zheng, H. Zhou, T.-B. Song, C.-C. Chen, P.S. Weiss, G. Li, Fused silver nanowires with metal oxide nanoparticles and organic polymers for highly transparent conductors, ACS Nano 5 (2011) 9877-9882. doi: 10.1021/nn203576v

[107] S. Watcharotone, D.A. Dikin, S. Stankovich, R. Piner, I. Jung, G.H. Dommett, G. Evmenenko, S.-E. Wu, S.-F. Chen, C.-P. Liu, Graphene– silica composite thin films as transparent conductors, Nano Lett. 7 (2007) 1888-1892. doi: 10.1021/nl070477+

Materials for Solar Cell Technologies II
Materials Research Foundations **104** (2021) 114-133

Materials Research Forum LLC
https://doi.org/10.21741/9781644901410-5

Chapter 5

Simulation Models for Solar Photovoltaic Materials

M. Rizwan[1,*], Waheed S. Khan[2], A. Asma[3], A. Shehzadi[3]

[1] School of Physical Sciences, University of the Punjab, Lahore, Pakistan

[2] Nanobiotechnology Group, National Institute for Biotechnology and Genetic Engineering (NIBGE), Jhang Road, Faisalabad-38000, Pakistan

[3] Department of Physics, University of Gujrat, Hafiz Hayat Campus, Gujrat City, Pakistan

*Corresponding Author: rizwan.sps@pu.edu.pk

Abstract

Semiconducting materials have dominated the photovoltaic industry for a long time. The advancement in solar cell technology is significantly influenced by computer modelling, designing and simulations of the semiconductor materials used for the device operation. Different modelling techniques including one, two and three dimensional models had been employed to comprehend the device operation of solar cell and other electronic devices based on semiconductor materials such as silicon and gallium arsenide. The performance of computing power is increasing with the passage of time in order to improve modelling and designing of different semiconductor materials for solar cell devices. In this chapter, different reported semiconductor materials, their standard characteristics and basic history of modelling, standard models used in photovoltaic industry and principles of modelling such as carrier statistics, transitions, band structure and mobility are explained in detail. Different characteristics of semiconductor material like the carrier transportation, carrier statistics, band structure, and heavy doping effect and carrier generations are described with respect to material modelling.

Keywords

Semiconducting Material, Photovoltaic Industry, Band Structure, Carrier Generation, Modelling

Contents

1. Introduction

Accurate and precise modelling of semiconductor devices is a necessity to comprehend semiconductor phenomenon like generation of heat and movement of electrons, impact ionization to make the functions in electronic devices more impeccable. For high field phenomenon the standard drift-diffusion phenomenon cannot be employed since they do not use energy as a dynamic variable. Moreover for the implementation of semiconductors in photovoltaic industry and optoelectronic devices the interactions of EM radiation and carriers is necessary. The general models used for semiconductor modelling used are called hydro dynamical models but they have theoretical drawbacks. A detailed knowledge about the parameters used in solar cells is necessary to understand the device operation of solar cell. In the modern era of knowledge, to comprehend complete device operation of solar cell requires enhancement in the computing power for efficient device operation in accordance with modelling, simulation and optimisation. The characteristics of semiconductor materials are closely related to a solar cell device operation. The semiconductor material characteristics are carrier statistics, transport,

Materials for Solar Cell Technologies II   Materials Research Forum LLC
Materials Research Foundations **104** (2021) 114-133   https://doi.org/10.21741/9781644901410-5

optical properties and recombination methods [1]. The side effects in solar cell device operation are caused due to harmful effects of the radiation that are experimentally observed under heavy doping effects [2].

In solar cell operation a lot of attention is paid to the properties of silicon and germanium that are the basic semiconductor material used in solar cell for energy generation. Various semiconductor fundamental parameters are encountered in photovoltaic applications. The antireflection coating can be utilized in semiconductors by using refractive index information of semiconductors as shown in table 1 [3-7]. Commercially, various computer programs are utilized for the modelling of materials to be used in solar cell operation. Some computer programs for semiconductor modelling are given below[8].

- Personal Computer 1-Dimensional (PC1D) software was established in the Institution of higher education Australia by P.A. Basore and co-workers for solar cell modelling. This 1-Dsimulant is ordinary utilized by public of photovoltaic cell industry[9].

- ATLAS simulation software by SILVACO International is utilized for two and three dimensional modelling of solar cell. A luminous tool that is the computing method is utilized for the ray tracing of response of a solar cell to the light. The monochromatic or multi-spectral sources of light are also utilized in this software[9].

- Another modelling method is MEDICI, it is a 2-D distribution of concentration of carriers and potential of the semiconductor, proposed by TMS (technology modelling associates). Photo generation can also be computed using this model. In a semiconductor device the potential and carrier concentration are computed through modelling technology that associates models in two-dimension via MEDIC1. The photon generation can be attributed to multi-tasking spectral sources, which encompass the modern application of optical device [10].

The accessibility of concentration of charge carrier is dealt with certain computing programs. In general, the modelling of solar cells or the modelling of semiconductor devices basically deals with the numerical techniques which can initiates the main principles of discussion [11].

Materials for Solar Cell Technologies II   Materials Research Forum LLC
Materials Research Foundations **104** (2021) 114-133   https://doi.org/10.21741/9781644901410-5

Table 1   The refractive index for common materials [3-7].

Material	Refractive index at $\lambda = 0.589\ \mu$m	ρ_p (g cm^{-3})
Water	1.33–0i	1.00
Silica	1.55–0i	2.66
Ammonium Sulfate	1.53–0i	1.76
Urban aerosol (avg.)	1.56–0.087i	1.7
Hazy aerosol (avg.)	1.96–0.66i	2.0
Clean aerosol (avg.)	1.53–0.015i	1.68

2.   History of semiconductor modelling

The developments in the past two decades in the semiconductor industry have led to great interest in material modelling. The quest behind to understand the device operation of silicon based simple devices to high integrated circuits has increased the importance of modelling in the device technology. With each passing day the size of electronic devices is decreasing, while at the same time the structure is getting more complicated. The miniaturization of devices results in complicating the measurement process. Modelling of devices not only leads to better understanding of device operation but also the new device structures. In semiconductor devices, two types of modelling is possible, physical and equivalent models. Physical models explain the physics behind the device operation and equivalent models explain the circuitry. The equivalent models are less practical due to the complexity while the physical models are very significant to comprehend the application of devices.

Physical models are realized through solution of semiconductor equations such as transport equations. Most models utilize the bulk carrier transport equations. The interest in material modelling started about 20 years ago. With time this field has become a well-established industry. Before the availability of wide spread computers, analytical tools were used. Shockley first provided a linear approximation tool that provides basic information regarding device operation but does not give any information regarding optimization. Thus, numerical simulations were introduced to encompass all parameters of device. In the early days, only one dimensional device modelling was used. In 1964, a one dimensional numerical model was used for GaAs.

Around 1969, two dimensional models came on the horizon. These models were used in BJT devices. These two dimensional models however weren't able to explain the 3 dimensional phenomenon in very small devices. Major solid state devices involve three dimensional phenomena such as hot electron generation. Monte Carlo technique for 3-dimensional modelling was introduced in 1966.This technique is used in Si, InP based devices. With quantum based semiconductor devices, new models are required [12]. All numerical simulations are based on the following principles.

3.   Semiconductor band structure

The operation of solar cells have fundamental importance and the main properties of semiconducting materials depends on the band gap which is the difference in valance band and conduction band energy and the structure of crystal is dependent on k-wave vector. The band gap energy and wave vector k is the main parameters that vary with temperature and pressure. The variation in the energy of band gap because of the temperature factor is mathematically designated by Varshni [13],

$$E_g(T) = E_{g0} - \frac{\alpha T^2}{T + \beta} \tag{1}$$

Where T, alpha and beta are the parameters, in which T is absolute temperature. For example for silicon the band gap energy was calculated 1.17eV and constants values are 4.730 and 636 respectively.

The optical transitions in between the band gap produce current through the solar cells. There are different kinds of transition in which two are mentioned here [14]. One is direct transition that occurs, when the net momentum is null from ion-electron transition then that transition is direct, another case is indirect transition in which addition of pair of electron-ion is considered [14, 15]. In indirect transition the phonon is required for the complete transition that is due to thermal quantum lattice vibration.

Semiconductors are of two kinds:

3.1   Direct band gap semiconductors

In direct band gap, when the top of valance band and bottom of conduction band occur at the same k-point of first Brillouin zone, such semiconductors are characterized as direct band gap semiconductors.

3.2   Indirect band gap semiconductor

In case of indirect band gap, the maxima of valance band and the minima of conduction band occur at the different k-point of first Brillouin zone [16], such semiconductors are called as indirect band gap semiconductors.

The direct band gap semi-conductor is stronger than indirect band gap [17] and is considered in case of optical absorption, for example the silicon and gallium arsenide etc. as shown in Figure 1 [18].

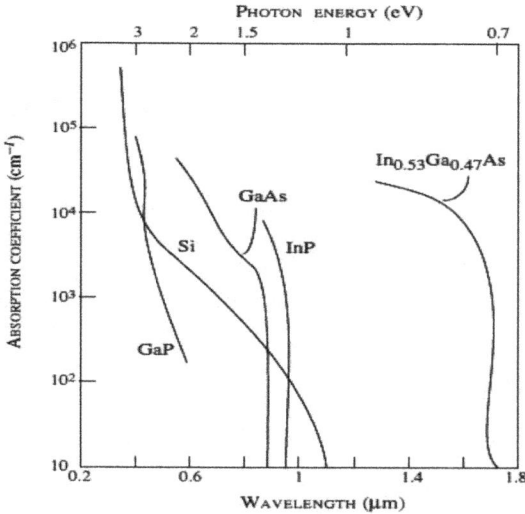

Figure 1 For Si, GaAs, InP and InGaAs the optical absorption. Reprinted from Ref. [18] under open access conditions.

4.   Semiconductor's carrier statistics

In thermal equilibrium, the two factors, temperature and the electrochemical potential represented by E_f is constant for a given device [19]. The concentration of electron is n and concentration of hole is represented by the p and the product of these concentrations is not dependent on substitution concentration. The law of mass action given by [20],

$$\text{np} = n_i{}^2 = N_c N_v \exp\left(-\frac{E_g}{K_B T}\right) \tag{2}$$

where T is the absolute temperature ,k is Boltzmann constant, ni is the intrinsic concentration of semiconductor of   electron and N_c, N_v are the effective density of conduction and valance band [21].

Materials Research Forum LLC
https://doi.org/10.21741/9781644901410-5

The effective density of states are $N_c N_v$ are given by,

$$N_c = 2 \left(\frac{2\pi m_e k_B T}{h^2}\right)^{3/2}$$

And $\hspace{8cm}$ (3)

$$N_v = 2 \left(\frac{2\pi m_h k_B T}{h^2}\right)^{3/2}$$

Here, m_e and m_h are the effective masses of electron and holes density of states and h is the planks constant.

The product of electron hole concentration does not obey for operation of solar cell from the equation 2. Differences may lead to transition of electron between different bands or quantum sates that will lead to gradients of the electrochemical potential and temperature and current also runs [17]. Then suitable kinds of carriers are explained for each energy band level to represent by a quasi-Fermi level. However, the quasi-Fermi level of concentration of electron in conduction band may be examined by E_{Fn} ,the holes concentration in valance band by E_{Fp} [22] .

It is important to explain the potential of electron and holes as given by,

$$E_{Fn} = -q\varphi_n$$

$\hspace{11cm}$ (4)

$$E_{Fp} = -q\varphi_p$$

The electron and holes carrier concentration terminologies are explained in equation 5 [23-30]. The solar cell operation does obey the above equation in order to explain the quasi-Fermi level of potentials of electron and hole [31].

$$N_c \exp\left(\frac{E_{Fn}-E_c}{K_B T}\right) \quad n_i \exp\left(\frac{q(\psi - \varphi_n)}{K_B T}\right) \quad N_c F_{1/2} \exp\left(\frac{E_{Fn}-E_c}{K_B T}\right)$$

$$N_v \exp\left(\frac{E_v - E_{Fp}}{K_B T}\right) \quad n_i \exp\left(\frac{q(\varphi_p - \psi)}{K_B T}\right) \quad N_v F_{1/2}\exp\left(\frac{E_v - E_{Fp}}{K_B T}\right) \hspace{1cm} (5)$$

5.   Carrier mobility

The current densities of electron and hole are given by [27],

$$j_n = q\mu_n n\varepsilon + qD\nabla_n n$$
$$j_p = q\mu_p n\varepsilon - qD\nabla_p n \qquad (6)$$

In equation 6,

μ_n= The mobility of electrons

μ_p= The mobility of holes

n= Concentration of electrons

p=Hole-concentration

D= Diffusion current for holes and electrons

And

ε= Electric field.

The first terms in equation 6 is due to electric fieldε, and second is accountable for diffusion. The drift mobility is ratio of drift velocity by field strength in above equation that explains the weak field. Electron and holes are charge carriers and dependence of impurity substitution on charge carriers is a function of temperature as in form of lesser or higher carriers. If we increase the temperature then the charge carrier concentration also varies according to situation. The Caughey-Thomas [32] is representing the empirical dependencies for case of silicon is given by,

$$\mu = \mu_{min} + \frac{\mu_0}{1+\left(\frac{N}{N_{ref}}\right)^\alpha} \qquad (7)$$

The different charge carrier values are in majority and minority as explained by the various parameters. For example the silicon, the majority carrier mobility and minority carrier mobility values are reported in literature [23, 33-35].

At high concentration, the effect of lattice on the scattering, impurity scattering, carrier-carrier scattering is significant [36].

When the carrier velocity reaches to saturation, there is a low current flow in same direction of electric field, and thus the mobility of carriers accelerates. The mean velocity that is the function of field is described as,

$$v = \frac{\mu_{lf}}{\left(1+\left(\frac{\mu_{lf}\varepsilon}{v_{sat}}\right)^{\beta}\right)^{1/\beta}} \qquad (8)$$

Here ß is parameter of carrier mobility as a low field, for electron its value is 2 and for holes equal to 1, v_{sat} is the velocity at saturation point same for both holes and electrons [37-39].
The saturation velocity is given by,

$$v_{sat} = \frac{2.4\times10^7}{1+0.8e^{\frac{T}{600}}} \qquad (9)$$

In case of gallium arsenide, the information collected for mobility of majority carriers is related to some literature review in [35], the data given is more difficult for numerical or empirical analysis. The mobility of minority carrier as a function of impurities concentration as shown in Figure 2 [40] and Figure 03 [41].

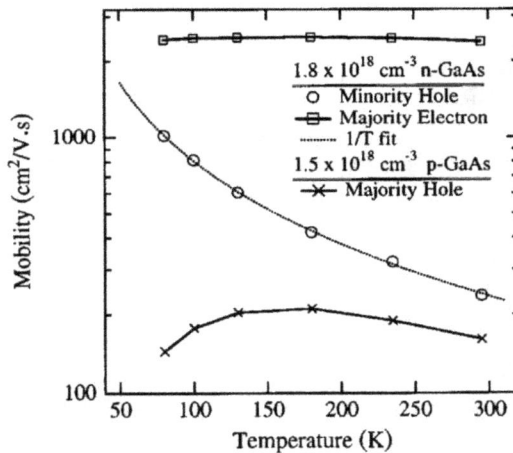

Figure 2 Minority-carrier mobility as a function of temperature [39]. Reprinted from ref. [40] copyright (1995), with permission from AIP Publishing.

Materials for Solar Cell Technologies II   Materials Research Forum LLC
Materials Research Foundations **104** (2021) 114-133   https://doi.org/10.21741/9781644901410-5

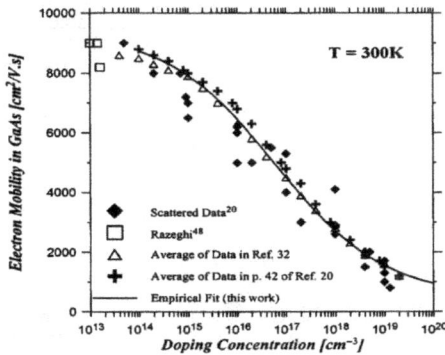

Figure 3 Mobility dependence on the doping concentration for gallium arsenide, [41].
Reprinted from ref. [41], copyright (2000), with permission from AIP Publishing.

The silicon and gallium arsenide carrier mobility are a function of electric field and the velocity is such and is given by as follows [34],

$$\mu_n = \frac{\mu_{lf} + v_{sat}\left(\frac{\varepsilon}{\varepsilon_0}\right)^{\beta}}{\left(1 + \left(\frac{\varepsilon}{\varepsilon_0}\right)^{\beta}\right)} \tag{10}$$

Where μ_{lf} is represented by the mobility of low field carriers, ε_0 is equal to 4×10^3 V/cm and the parameter beta is equal to 4 for electrons and 1 for holes [1, 33].

The v_{sat} is for silicon and gallium arsenide is given by [42],

$$v_{sat} = 11.3 \times 10^6 - 1.2 \times 10^4 T \tag{11}$$

Here, T is represented by absolute temperature.

Indium phosphide the majority carrier mobility is reported in figure 4.

123

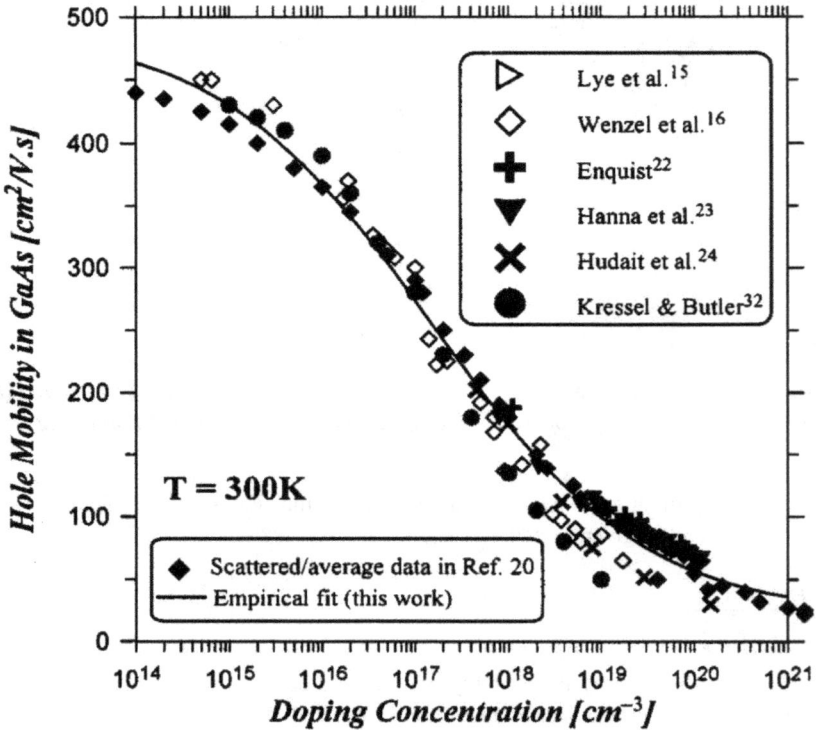

Figure 4 Mobility dependence on the doping concentration for indium phosphide. Reprinted from ref. [41] copyright (2000), with permission from AIP Publishing.

6.   Carrier generations by optical absorption

6.1   Band-to-band transitions

In solar cells main function of carrier generation is the light absorption. Consider for the case of planar slab, the photon which move to semiconductor that produces the pairs of electron and hole, $g(x)\delta x$, of thin layer of slab from depth $x \to x + \delta x$. The generation Function $g(x)$ is given by,

$$g(x) = \alpha(\lambda)exp\{-\alpha(\lambda)x\}$$

Here, $\alpha(\lambda)$ is represented by absorption coefficient, absorption coefficient as a function of energy is given in figure 5 [43]. The g(x) is a function define by the ratio of generation rate G to volume that's define by the above equation

Generation Function =g/A

In above equation A indicated the sample area.

Rajkananand co-workers [19] reported that a fundamental formula for the case of silicon is described by the absorption coefficient,

$$\alpha(T) = \sum_{\substack{i=1,2 \\ j=1,2}} C_i A_j \left[\frac{\{hv - E_{gj}(T) + E_{pi}\}^2}{\{exp(E_{pi}(KT)-1\}} + \frac{\{hv - E_{gj}(T) - E_{pi}\}^2}{\{1 - exp(E_{pi}(KT))\}} \right] + A_d \{hv - E_{gd}(t)\}^{1/2} \qquad (12)$$

Here, hv show the photon energy, $E_{g1}(0) = 1.1557eV$, $E_{g2}(0) = 2.5eV$ and $E_{gd}(0) = 3.2eV$ the energies shows that the lowest indirect band gap and lowest direct band gap, respectively. Now, represented by the transverse optical and transverse acoustic phonons of the Debye frequencies $E_{p1} = 1.827 \times 10^{-2}eV$ and $E_{p2} = 5.773 \times 10^{-2}eV$, and $C_1 = 5.5$ & $C_2 = 4.0$ and $A_1 = 3.231 \times 10^2 cm^{-1}eV^{-2}$, $A_2 = 7.237 \times 10^3 cm^{-1}eV^{-2}$ and $A_1 = 1.052 \times 10^6 cm^{-1}eV^{-2}$. The band gap is a function of temperature, for example if we change the one parameter second is automatically changed according to our desired requirements. These parameters $\alpha = 7.021 \times 10^{-4}$, eV/k^2 and $\beta = 1180k$ are described by the three band gap energies E_{g1}, E_{g2} and E_{gd}.

Materials for Solar Cell Technologies II   Materials Research Forum LLC
Materials Research Foundations **104** (2021) 114-133   https://doi.org/10.21741/9781644901410-5

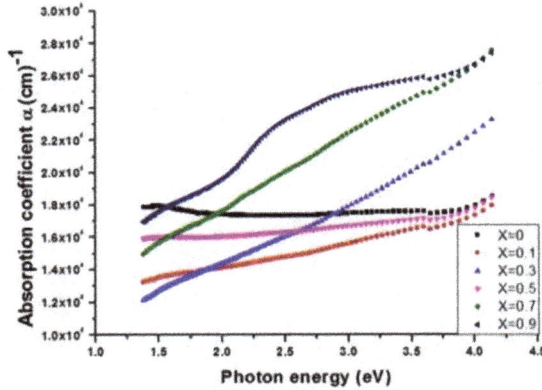

Figure 5 For solar cell the absorption coefficient as a function of energy[43]. Reprinted from ref. [43] under the terms of Creative Commons License

6.2   Free-carrier absorption

Electron transition from an initial state to final state upon absorption of photon can happen in a region in the same band due to two parameters one is impurity and other is because of strong interactions within the band. As a result of such transition current is generated by the band-band transition rather than electron hole pair that is produced due to free carrier absorption. The photon energies might be equal to band gap for the case of free carrier absorption rather than absorption coefficient and could be included in form of exponential equation.

The region of high impurity or doping concentration depend on where the band edge is residing in various form of phenomena's.

For the free-carrier concentration uses the PCID model to calculate absorption coefficient and computes on the experimental data:

$$\alpha_{FC} = K_1 \, n \, \lambda^a \; + \; K_2 \, p \, \lambda^b \tag{13}$$

Where, these parameters are described in a, b, K_1 and K_2 in nanometre dimensions. For various parameters such as o mobility of electrons and holes are reported. The free carrier absorption co-efficient a, b, k_1 and k2 have constant values. For example silicon, gallium arsenide and indium phosphide were reported with their different constant values.

6.3   Heavy doping effects

The electronic properties of semiconducting material exhibit unless impurity density is very high. The depiction of electronic properties is not dependent on the interaction of electron to the isolated impurities. Two impurity bands are formed when the doping concentration are enhanced, one is impurity band and the other is separate from the coulomb wave function due to other doping effects on impurity band [44]. Both bands are combined together and form a merged band that describe the localised states. To describe the semiconductor device operation and the optical parameters, the reasonable interaction in between the electron-electron that gives exchange energy and co-relation energy are discussed.

Band anisotropy and effective masses are utilized to calculate more accurate results for the practical purpose to explain the complex phenomenon of band gap narrowing. Absorption at low temperature data is utilized for the optical measurement of the device operation. The band gap narrowing is represented by the symbol ΔE_g . For the case of n-type material is given by,

$$\Delta E_g = KT \ln \frac{N_D p_0}{n_i^2} \tag{14}$$

Here, p_0 is represented by the concentration of minority holes and N_D represent the doping donor concentration.

From literature review, the values can be compared with experimental work that has been computed from the different phenomena's. Sample is prepared using different material to study the band gap narrowing by the Jain and co-workers [45]. Klaasenet co-workers [46] and its co-worker are working on the device modelling of semiconductor based silicon. By comparing the theoretical corrected values with the experimental data from literature review fives a new estimated value of intrinsic free carrier concentration of the n and p-type of narrowing band as described by Klaasenet and co-workers [46] to single expression,

$$\Delta E_g(meV) = 6.92 \left[\ln \frac{N_{dop}}{1.3 \times 10^{17}} + \sqrt{\left(\ln \frac{N_{dop}}{1.3 \times 10^{17}} \right)^2 + 0.5} \right] \tag{15}$$

Here the doping concentration is shown by N_{dop} .

Lundstromet.al[40] suggested the formula consists of doping concentration for gallium arsenide is empirical data:

Materials for Solar Cell Technologies II
Materials Research Foundations **104** (2021) 114-133

Materials Research Forum LLC
https://doi.org/10.21741/9781644901410-5

$$\Delta E_g = A N_{dop}^{1/3} + K_B T \ln\{F_{1/2} (E_F)\} - E_F \tag{16}$$

Where, the Fermi energy is represented by E_F. The function $F_{1/2}$ given by,

$$A = \begin{cases} 2.55 \times 10^{-8} eV & (p-GaAs) \\ 3.23 \times 10^{-8} eV & (n-GaAs) \end{cases} \tag{17}$$

7.   Modelling simulation and analysis

The progress in Nano/microelectronic is rising as in comparison to the fundamental model TCAD, because one is high cost and other is developing long range order of simulation, optimisation and modelling of geometry. Firstly, the real part of components can be classified individually for the whole explanation, and then examine the element of that system at the end and observe the connection between them.

Simulation is a process in which we can solve the problem through some systems of operation without fabrication. And the theoretically simulation guess is more precise, if we accept the idea of approximation. Modelling is basically defined by the process or device, to find out the way to solve the problem, the analysis of the different conception about geometry that can't be measured or calculated exactly. Basic requirements are simulation, analysis and geometry optimisation for modelling. There are two models, one is used for the curves fitting such as known simulation models that are easier than other models for analysis such as described by their physical aspects in qualitative way. The most common example is Monte Carlo method, one of the applications of the models. It is a simulation method use for various functioning in a system. Using this model, we must be aware of the constraints of the appliances for achieving more accurate results. And we must be comparing the results from literature review in theoretical and experimental papers.

Field of electronic devices is progressing, by designing the geometry for various operations using different parameters to obtain more accurate results during manufacturing for tests. In modern era, the simulation process must be much better to obtain the number of iterations. The simulation and analysis gives 40% success roughly. This estimation clearly depends on the model that is used for simulation. Due to scraps and tests, the uncertainty in parameters can't be removed. The quantum effects prevail, due to the scaling factor used is too large for the model. The theoretical resources are less costly as compared to experimental, which sustain the set expectations. Numerical modelling is utilized for large number of calculations.

Materials for Solar Cell Technologies II   Materials Research Forum LLC
Materials Research Foundations **104** (2021) 114-133   https://doi.org/10.21741/9781644901410-5

Conclusion

It is established that among all the techniques, MEDICI is the most efficient technique for solar cell device operation. Since this three dimensional technique encompasses all aspects of device operation and is ideal for devices based on quantum aspects and also for the miniaturization of electronic devices. The decrement in the size of transistor is a huge recurring demand in semiconductor industry, which increase demand in decrement. The material growth for low dimensional devices has been increased highly. We have shown that some models and techniques for computers modelling and simulation agreed well with theoretical and experimental results. Initially two types of models were used, physical and equivalent, but only physical models survived the industry due to their significance and importance in photovoltaic industry. It is essential to apprehend the physics behind the semiconductor devices for their good performance in application by low dimensional material of simulation and modelling using physical models. These results can be practically applied to electronic devices such as circuit-level simulation for various device models such as MOSFETS, BJT-devices. The cell generated photocurrent, current-voltage graph, carrier statistics and series resistance are used for different methods to estimate the simulations. In upcoming years, various advanced modelling techniques such as those based on quantum mechanics are expected that will lead to revolutionary changes in photovoltaic industry.

Acknowledgements

Authors are immensely grateful to Department of Physics, University of Gujrat, Gujrat-50700, Pakistan and National Institute for Biotechnology & Genetic Engineering (NIBGE), Faisalabad-38000, and Pakistan for providing full support and healthy environment for such type of writing work.

References

[1] T. Markvart, L. Castaner, Semiconductor materials and modelling, in: A. McEvoy, T. Markvart L. Castaner (Eds.) Practical handbook of photovoltaics: Fundamentals and applications Academic Press (2011). https://doi.org/10.1016/B978-0-12-385934-1.00002-7

[2] U. Rau, H.W. Schock, Electronic properties of Cu (In, Ga) Se$_2$ heterojunction solar cells–recent achievements, current understanding, and future challenges, App. Phy. A 69 (1999) 131-147. https://doi.org/10.1007/s003390050984

[3] S. Swirhun, Y.H. Kwark, R. Swanson, Measurement of electron lifetime, electron mobility and band-gap narrowing in heavily doped p-type silicon, 1986 International

Electron Devices Meeting, IEEE, 1986, pp. 24-27.
https://doi.org/10.1109/IEDM.1986.191101

[4] J. Del Alamo, S. Swirhun, R. Swanson, Simultaneous measurement of hole lifetime, hole mobility and bandgap narrowing in heavily doped n-type silicon, 1985 International Electron Devices Meeting, IEEE, 1985, pp. 290-293. https://doi.org/10.1109/IEDM.1985.190954

[5] S. Sze, Physics of semiconductor devices, John Wiley, New York NY (1981) 122-129.

[6] P. Garcia-Nieto, Study of visibility degradation due to coagulation, condensation, and gravitational settling of the atmospheric aerosol, Aerosol. Sci. Tech 36 (2002) 814-827. https://doi.org/10.1080/02786820290092069

[7] A. Elfakir, T. Tlemçani, E. Benamar, A. Belayachi, E. Gutierrez-Berasategui, G. Schmerber, M. Balestrieri, S. Colis, A. Slaoui, A. Dinia, Structural, electrical and optical properties of sprayed Nd–F codoped ZnO thin films, J. Sol-Gel. Sci. Techn73 (2015) 557-562. https://doi.org/10.1007/s10971-014-3518-y

[8] S. Selberherr, Analysis and simulation of semiconductor devices, Springer Science & Business Media (2012).

[9] T.P. Pearsall, Properties, processing and applications of indium phosphide, Institution of Electrical Engineers, 2000.

[10] M. Shur, Physics of Semiconductor Devices, Prentice Hall, Inc., Englewood Cliffs, New Jersey (1990) 680.

[11] L.L. Kazmerski, S. Wagner, Cu-ternary chalcopyrite, Solar cells, in: T.J. Coutts, J.D. Meakin (Eds.), Current Topics in Photovoltaics, Academic Press, Orlando (1985) 41.

[12] C.M. Snowden, Semiconductor device modelling, Rep. Prog. Phys 48 (1985) 223. https://doi.org/10.1088/0034-4885/48/2/002

[13] Y.P. Varshni, Temperature dependence of the energy gap in semiconductors, physica 34 (1967) 149-154. https://doi.org/10.1016/0031-8914(67)90062-6

[14] M. Littlejohn, J. Hauser, T. Glisson, Velocity field characteristics of GaAs with Γ^c $_6$-L ^{c} $_6$-X ^{c} $_6$ conduction-band ordering, J. Appl. Phys 48 (1977) 4587-4590. https://doi.org/10.1063/1.323516

[15] E. D. Palik, , Handbook of optical constants of solids, Academic Press Handbook Series (1985).

Materials for Solar Cell Technologies II   Materials Research Forum LLC
Materials Research Foundations **104** (2021) 114-133   https://doi.org/10.21741/9781644901410-5

[16] P.J. Timans, The thermal radiative properties of semiconductors, in: F. Roozeboom (Ed.) Advances in rapid thermal and integrated processing, Springer (1996), pp. 35-101. https://doi.org/10.1007/978-94-015-8711-2_2

[17] Z. Liang, A. Nardes, D. Wang, J.J. Berry, B.A. Gregg, Defect engineering in π-conjugated polymers, Chem. Mater. 21 (2009) 4914-4919. https://doi.org/10.1021/cm902031n

[18] D. Let, A. Stancu, V.G. Cimpoca, Study over optical absorption and emission in semiconductors, J. Sci. Arts.

[19] K. Rajkanan, R. Singh, J. Shewchun, Absorption coefficient of silicon for solar cell calculations, Solid State Electron. 22 (1979) 793-795. https://doi.org/10.1016/0038-1101(79)90128-X

[20] P. Schmid, Optical absorption in heavily doped silicon, Phys. Rev. B 23 (1981) 5531. https://doi.org/10.1103/PhysRevB.23.5531

[21] A. Ramdas, S. Rodriguez, Spectroscopy of the solid-state analogues of the hydrogen atom: donors and acceptors in semiconductors, Rep. Prog. Phys. 44 (1981) 1297. https://doi.org/10.1088/0034-4885/44/12/002

[22] W. Van Roosbroeck, W. Shockley, Photon-radiative recombination of electrons and holes in germanium, Phys.Rev. B 94 (1954) 1558. https://doi.org/10.1103/PhysRev.94.1558

[23] R. King, R. Sinton, R. Swanson, Studies of diffused phosphorus emitters: saturation current, surface recombination velocity, and quantum efficiency, IEEE T. Electron. Dev 37 (1990) 365-371. https://doi.org/10.1109/16.46368

[24] H.J. Hovel, R.K. Williardson, A.C. Beer, Semiconductors and semimetals. Volume 11. Solar cells, Academic Press, New York (1975). https://doi.org/10.1063/1.3024511

[25] F. Seitz, Displacement of atoms during irradiation, Solid. State. Phys. 2 (1956) 307-442.

[26] G. Kinchin, R. Pease, The displacement of atoms in solids by radiation, Rep. Prog. Phys18 (1955) 1. https://doi.org/10.1088/0034-4885/18/1/301

[27] B. Anspaugh, GaAs solar cell radiation handbook, JPL Publication, California, 1996.

[28] H. Tada, J. Carter Jr, B. Anspaugh, R. Downing, Solar cell radiation handbook, JPL Publication, California, 1982.

[29] T. Coutts, M. Yamaguchi, Indium phosphide-based solar cells: A critical review of their fabrication performance and operation, Current topics in photovoltaics 3 (1988).

Materials Research Forum LLC
https://doi.org/10.21741/9781644901410-5

[30] V.M. Andreev, V.A. Grilikhes, V.D. Rumiantsev, Photovoltaic conversion of concentrated sunlight, John Wiley1997.

[31] B. Garcia, J. Martinez, J. Piqueras, Laser melting of GaAs covered with thin metal layers, Appl. Phys. A 51 (1990) 437-445. https://doi.org/10.1007/BF00348387

[32] D. Caughey, R. Thomas, Carrier mobilities in silicon empirically related to doping and field, P. IEEE 55 (1967) 2192-2193. https://doi.org/10.1109/PROC.1967.6123

[33] R.G. Downing, J.R. Carter Jr., J.M. Denney, The energy dependence of electron damage in silicon, Proc. 4th IEEE, Photovoltaic Specialists Conf., 1 (1964) A-5-1.

[34] W. Rosenzweig, Diffusion length measurement by means of ionizing radiation, BSTJ 41 (1962) 1573-1588. https://doi.org/10.1002/j.1538-7305.1962.tb03995.x

[35] A. Meulenberg, Damage in silicon solar cells from 2 to 155 MeV protons, Photovoltaic Specialists Conference, 10 th, Palo Alto, Calif, 1974, pp. 359-365.

[36] T. Markvart, Radiation damage in solar cells, J. Mater. Sci-Mater El.1 (1990) 1-12. https://doi.org/10.1007/BF00716008

[37] M. Yamaguchi, K. Ando, Mechanism for radiation resistance of InP solar cells, J. Appl. Phys63 (1988) 5555-5562. https://doi.org/10.1063/1.340332

[38] C. Travier, Reviw of microwave guns, Part. Accel. 36 (1991) 33-74. https://doi.org/10.1002/ecja.4410741204

[39] M.L. Lovejoy, M.R. Melloch, M.S. Lundstrom, Temperature dependence of minority and majority carrier mobilities in degenerately doped GaAs, App.Phys.Lett 67 (1995) 1101-1103. https://doi.org/10.1063/1.114974

[40] M. Lundstrom, E. Harmon, M. Melloch, Effective bandgap narrowing in doped GaAs, EMIS datareviews series 16 (1996) 186-189.

[41] M. Sotoodeh, A. Khalid, A. Rezazadeh, Empirical low-field mobility model for III–V compounds applicable in device simulation codes, J. Appl. Phys87 (2000) 2890-2900. https://doi.org/10.1063/1.372274

[42] W. Rosenzweig, F. Smits, W. Brown, Energy dependence of proton irradiation damage in silicon, J. Appl. Phys35 (1964) 2707-2711. https://doi.org/10.1063/1.1713827

[43] S.A. Hussain, G.J.A. Sada, The effect of cu-doping on the optical properties of znfe2o4 films prepared by chemical spray pyrolysis method, world (Drude) 3 (2009) 4.

[44] W. Shockley, W. Read Jr, Statistics of the recombinations of holes and electrons, Phys.Rev 87 (1952) 835. https://doi.org/10.1103/PhysRev.87.835

[45] S. Jain, D. Hirst, J. O'sullivan, Gold nanoparticles as novel agents for cancer therapy, Br. J. Radiol. 85 (2012) 101-113. https://doi.org/10.1259/bjr/59448833

[46] D. Klaassen, J. Slotboom, H. De Graaff, Unified apparent bandgap narrowing in n- and p-type silicon, Solid-State Electron. 35 (1992) 125-129. https://doi.org/10.1016/0038-1101(92)90051-D

Materials for Solar Cell Technologies II
Materials Research Foundations 104 (2021) 134-148

Materials Research Forum LLC
https://doi.org/10.21741/9781644901410-6

Chapter 6

Solar Energy and its Multiple Applications

David A. Wood[1*], Mohammad Reza Rahimpour[2]

[1]DWA Energy Limited, Bassingham, Lincoln, LN5 9JP, United Kingdom

[2]Department of Chemical Engineering, Shiraz University, Shiraz, 71345, Iran

* dw@dwasolutions.com

Abstract

Solar energy is commercially exploited to provide benefits in the form of various products and capabilities applying a range of technologies. Electricity generation is achieved either directly from photovoltaic cells made of various materials or indirectly through the steam production from concentrating solar thermal systems. Whereas solar thermal power generation requires large scale plants, photovoltaic systems can be large or small in scale and building integrated, if required. Both types of generation can be standalone or connected to power grids. Solar energy is also extensively used for water and space heating, cooling and drying purposes. It can also be stored and/or transformed into a range of clean fuels and contributes energy to the manufacture of various energy-intensive products. The research into the artificial photosynthetic synthesis of biofuels although encouraging is, however, yet to be achieved commercially exploited on a large scale. Much scope remains for innovative technology breakthroughs to further improve the efficiency and uptake of all the solar energy technologies currently exploited or under investigation. Policy frameworks, renewable portfolio standards, feed-in tariffs and net-metering play an important and ongoing role in promoting the uptake of photovoltaics in particular.

Keywords

Solar Energy, Photovoltaic, Concentrating Solar Thermal, Thermochemical Transformations, Evaporative Cooling, Net Metering

Contents

1. Introduction

To date, humanity has experienced miscellaneous improvements to its quality of life and gained access to resources that have expanded its capabilities. Many of these benefits have resulted from developments and refinement associated with innovative technologies. The primary aim of seeking and exploiting new technologies is to enhance prosperity while improving the quality of human life. For instance, new technologies that enhance energy resource recovery [1-6] and /or improve efficiency in a sustainable manner constitute beneficial advances for human existence. The diverse and rapid development of innovative technologies associated with the exploitation of solar energy has the potential to provide a significant environmentally friendly and sustainable portion of future energy supply.

The sun provides about 99.9% of the world's total energy [7,8]. Nuclear fusion reactions occurring within the sun are responsible for the generation of energy within the sun. These reactions have been operating for billions of years and will likely persist for billions of years into the future. The sun releases its energy at a mass-energy conversion rate of 4.26 million metric tons per second, which equates to 3.846×10^{26} watts. However, the earth receives just 0.000000045% of this power [9]. This is because the energy emitted by the sun is spread over the area of a sphere with a radius of about 93,000,000 miles as it impacts the earth.

2.   Insight to solar energy in terms of its potential and challenges

Solar energy possesses a wide range of benefits compared with other energy sources. It is incident and available at some point each day across the entire earth's surface but with different intensity at each location. This means that solar energy as a resource is in plentiful supply, readily available to exploit and sustainable for the foreseeable future. Solar energy has the potential to be transformed into several different and exploitable forms of energy. This means that it can be flexibly configured for beneficial uses in a wide range of applications at many locations around the globe. These transformations are environmentally friendly as they do not involve the formation of pollutants such as greenhouse gases. Conversion of solar radiation at low concentrations can be usefully employed in photovoltaic or solar thermal receivers that are suitable to install on the rooves of urban buildings.

Solar energy does have some disadvantages as well, as being intermittent and not immediately available during night-time, it is more dispersed as an energy source than most other sources, such as fossil fuels. Consequently, solar power facilities require wide surface areas to gather large amounts of solar energy when it is available. The spectrum and severity of sunlight are different at each point on the earth's surface which substantially affects solar energy availability. These characteristics of sunlight are themselves functions of the dimensions of gaseous material that light from the sun must penetrate as it traverses the atmosphere to reach the earth's surface. Additionally, aerosols, clouds, dust, and smoke can play a negative role by diminishing the intensity of light energy arriving at the planet's surface. This means that some geographic locations, i.e., those with a high number of hours with clear skies, low atmospheric pollution levels and relatively low latitudes, can generate more solar power per unit surface area.

Another application of solar energy includes utilizing solar energy as a complementary source of energy to assist fossil fuels to provide more energy in hybrid power generation systems [10]. In terms of economic benefits, hybrid solar power facilities can be adapted to exploit some of the available sites and infrastructure associated fossil-fuel plants to provide incremental power at relatively low cost of supply. Optimal land allocation needs to be carefully planned for large non-urban solar power stations such that they are integrated, accessible and adequately backed-up by more continuously available power supply and adequately synchronized for smooth entry to the power transmission /

Materials for Solar Cell Technologies II   Materials Research Forum LLC
Materials Research Foundations **104** (2021) 134-148   https://doi.org/10.21741/9781644901410-6

distribution grid. Various, commercially viable solar energy conversion technologies are now readily available and selection among them requires careful consideration.

Most solar energy transformation technologies currently involve higher levelized cost of supply than most fossil fuels, although the cost-of supply trend for non-concentrating photovoltaics has reduced rapidly in recent years [11]. That trend is attracting more capital globally into solar energy projects [12]. Large-scale solar energy systems do require substantial capital investments and require careful quality control, project budgeting, planning and management in order to deliver cost-competitive facilities.

3   Advance utilization of solar energy

Procedures employed to gather solar energy radiation and transform it into readily useable forms of energy have been developed over many decades and innovations continue to improve efficiency. The products of these procedures include low-grade heat, high temperature steam, synthesis gas, hydrogen, ammonia, metals, and power. The technologies involved to produce such wide range of products are at varying levels of development and sophistication.

3.1   Photovoltaics configured to generate electricity on a large scale

Photovoltaic (PV) technology originated in the 1950s, was developed commercially during the 1990s and PV panels were first mass produced in 2000 [13]. Since the early 2000 global installed PVs capacity has grown almost exponentially and the technologies involved continue to be improved. Crystalline and amorphous silicon and cadmium telluride photovoltaics have become economically viable competitors to fossil fuels over the past decade. Many governments have welcomed and incentivized the developing photovoltaic integration into electricity networks. Indeed, subsidies and/or mandates that insist upon power system operators facilitating the preferential entry of solar, and other renewable power generated into power grids.

Thin-film photovoltaic cells offer an option to crystalline silicon cells. These are made of either amorphous silicon or metallic compounds able to convert light into electrical energy. Many wafer-based (polysilicon) photovoltaic cells are more expensive to manufacture, and because they are heavier than thin films can be more expensive to transport and set-up. Those cells made of thin-film materials involve a small fraction of the volume of components compared to rigid silicon photovoltaic set-ups. An added attraction is that thin film is able to contribute to roofing material, building cladding, transparent materials for window use and a wide range of applications requiring light flexible materials (e.g. foldable energy blankets, drones, road vehicles etc.). Photovoltaic cells exploiting thin-film materials are typically constructed by building up thin (semi-conductor) layers on to a solid (glass /stainless-steel) substrate. This process requires carefully controlled conditions and is typically performed in vacuum chambers. Laser scribing is required to isolate and electrically connect each electrical cell making up the thin-film photovoltaic module. The semiconductors used are often metallic, especially combinations of copper, indium, gallium, and selenium (CIGS) have received much investment, research and development over recent years. The four distinct semiconducting metals of a copper, indium, gallium, and selenium (CIGS) cell require careful deposition onto the substrate in order to function efficiently. This adds cost to the manufacturing process.

Moderately efficient CIGS cells involve a very small mass fraction of the silicon required for wafer-based cells but they remain more expensive to manufacture because of process complexity and materials costs. Technology breakthroughs are required to improve their energy conversion efficiency, which is typically between 10 to 15%, and reduce their manufacturing costs. Such improvements could expand the photovoltaic market share of thin film technologies from their current position of less than 20%. That market is currently dominated by silicon technologies with their higher energy conversion efficiency, typically from 15 to greater than 20%.

Large-scale commercial uptake of solar photovoltaics began in Japan, followed by Germany. Significant demand in those countries initiated mass production of photovoltaic cells which rapidly expanded to manufacturing centers in China and the United States as well, as the development of photovoltaic cell technologies to safely connect and integrate photovoltaic generation systems into formerly closed grids was also resolved. This has enabled very large-scale photovoltaic facilities (solar parks and farms) to be embedded in existing grids and deliver electricity at operational costs comparable to traditional electricity generation technologies. The generated electricity from such large-scale photovoltaic systems can now be for general grid supply and does not need to be dedicated to specific users [14]. The International Energy Agency's [IEA] Photovoltaic Power Systems Program [PVPS] [13] now follows miscellaneous set-up classifications in a wide range of countries. These systems are predominantly terrestrial [15]. Uptake of these systems has been extensive in Asia, Europe and North America. The main consumer markets for these large-scale photovoltaic developments have grown rapidly over the past decade in Asia and Africa. Scale and intermittency are the key hurdles that need to be overcome by photovoltaics to sustain continued growth in the global market share of power supply. In recent years for larger-scale photovoltaic generation plants it has become more common for the orientation of the solar cells to be continuously adjusted to face the sun across the course of daylight hours. Such tracking systems substantially increase the amount of power that can be generated.

3.2 Photovoltaics for the purpose of small-scale supply to residential consumers

Many electricity utilities still express some resistance to integrating photovoltaic solar power generation systems into their networks. This is not the case for small-scale, off-grid utilization of photovoltaics dedicated to individual residential and commercial buildings, and on the scale of small communities. Such systems, usually located on the building roofs, are able to conjoin to consumption meters and eliminate or significantly reduce consumption from the utility-controlled power networks. In some cases, such small-scale photovoltaic systems can be connected to the main power grids and export some of the power they generate to the grid and be paid a specified feed-in tariff for that supply. In terms of unit capital and operating costs, small-scale systems are substantially more expensive than utility-scale systems due to the economies of scale exploited by the latter. Nevertheless, these systems can be competitive in terms of cost of supply in the retail power sector. They can also improve security of supply to locations not well connected to the power networks and/or subject to frequent grid interruption. Where such systems are installed on a standalone basis, not connected to the grid systems, there setup costs are lower than those designed to benefit from feed-in tariffs. Small-scale solar facilities typically do not incorporate solar tracking capabilities due to cost issues but this may change as such systems evolve.

3.3 Solar thermal electricity generation

As the concentration of solar flux is a function of the surface area and geographic location, constructing large-scale concentrating solar thermal (CST) plants, as with photovoltaic technologies, requires large sites. They cannot therefore, be easily embedded within urban areas or other densely populated locations. For CST plants efficiency always requires mechanical tracking for the solar collectors continuously orienting them towards the sun.

Many installations utilizing CST technologies have been employed in USA and Spain. Indeed, the early history of CST technologies lead to greater uptake for large-scale power generation than photovoltaics [16]. However, the pace of growth of CST for large-scale plants has slowed relative to photovoltaics in recent years. However, several commercial-scale CST plants have also been built in Chile, Egypt, Morocco and the United Arab Emirates. China, India, and Australia have also developed mainly experimental and pilot CST plants. However, globally there is about 3.8 GW of CST capacity constructed with almost double that capacity under evaluation, planning and/or construction. The mining

Materials for Solar Cell Technologies II   Materials Research Forum LLC
Materials Research Foundations **104** (2021) 134-148   https://doi.org/10.21741/9781644901410-6

industry, particularly in Chile and in planning in Australia, are particularly attracted to CST power supplies in remote off-grid areas.

In terms of cost, CST possesses the advantage that it can be more easily connected to large-scale thermal energy storage systems than photovoltaics. Such thermal storage systems typically exploit latent heat or thermochemical conversion reactions to store energy for up to 12 hours or more smoothing the power generation outputs of CST systems across typical 24-hour periods involving day and night cycles. Integrated storage allows providing movable energy depending on demand. Large CST systems typically generate high temperatures used to raise steam and power steam turbines for electricity generation. It is also possible to develop hybrid systems that combine CST with combined cycle gas turbines, as has been done in Egypt and Morocco for example.

Concentration of solar energy is essential for CST systems to achieve high temperatures and more desirable electricity conversion efficiencies. There are various CST technologies developed and applied, including linear parabolic trough reflectors, power towers, Fresnel lenses, and concentrating dishes [17]. Linear focusing parabolic collectors are the most common among the earlier plants constructed. These tend to be associated with one-axis solar tracking and are capable of providing moderate temperatures up to about 400°C. Besides, paraboloidal dishes and centralized power towers which are associated with two-axis tracking provide greater solar energy concentration ratios and, consequently, resulting in greater temperatures. In a solar thermal power plant constructed using a central power tower, the light is reflected from an array of heliostats (moveable mirrors)towards the top of a tower. This involves a large number of heliostats arranged in concentric circles around the power tower each driven by a two-axis tracking system. The Ivanpah CST plant in the Mojave Desert California (377MW), commissioned in 2014, uses such a power tower system. That plant includes 300,000 mirrors and three 459-feet-tall towers. It provides enough electricity to power about 140,000 houses.

Currently, solar thermal generation of electricity remains much less economically beneficial in comparison with fossil-fueled power plants and photovoltaic plants in most locations. Consequently, it tends to be utilized in a location with clear-sky conditions and relatively high solar intensities and in more remote off-grid locations (e.g., to power remote mines in Chile). Technology breakthroughs and economies of scale have the potential to change this, particularly in the Middle East and parts of China.

3.4   Exploiting solar power to conduct thermochemical transformations

There are various energy storage technologies with the potential to decrease the intermittency constraints related to large-scale solar thermal plants. They include

Materials for Solar Cell Technologies II   Materials Research Forum LLC
Materials Research Foundations **104** (2021) 134-148   https://doi.org/10.21741/9781644901410-6

exploiting actual and latent heat to drive endothermic thermochemical processes [18]. Some of the energy storage technologies integrated with CST plants exploit phase changes in mineral salts at high temperatures. In other cases, the generated chemicals associated with thermochemical reactions can be transported, stored and/or directly introduced to reformer reactors where heat is ultimately released in a controlled manner from associated exothermic reactions.

It is also possible to utilize solar-generated heat to conduct endothermic industrial chemical conversion processes. For instance, reactions that alter the chemical content and or molecular structure of a petroleum product to enhance its calorific value and/or increase its hydrogen content. Also, reforming processes that generate hydrogen or synthesis gas from hydrocarbon feedstocks are endothermic and could use solar power as their energy source. To be effective such reactions need to be of large-scale capacity. Some pilot plants have so far been developed [19]. This choice of energy storage and/or chemical transformation differs from latent and actual heat storage processes because the energy can essentially be retained at ambient temperatures. Solar-produced fuels have the potential to provide considerable energy transformation capacity with relatively small investments required in infrastructure to produce them [19]. Other potential product manufacturing product transformation reactions that could be driven by solar energy are the production of lime, cement, ammonia and the smelting of metals. However, those products cannot be utilized for energy storage purposes. Utilizing solar energy to power these varied production processes reduces greenhouse gas emissions and, in the case of petroleum product production, limits the environmental footprint of certain refining and petrochemical processes. Involving solar power in selected refining and petrochemical processes could be a way of extending the life of petroleum resources by reducing their carbon footprints.

3.5 Heating water and space with solar energy

One relatively easier exploitation of solar energy is to generate hot water at various scales. Today, solar water heating systems are widely utilized around the world at small, residential scales up to large industrial scales. It represents a substantially exploited renewable energy technology [20].

Domestic hot water production exploiting solar thermal technologies is now commonly integrated with many buildings. It is estimated that some 18% of total energy use in the buildings in the USA, and 14% in the European Union is associated with hot water consumption [21]. Providing more of this resource from solar energy has the potential, therefore, to result in a significant reduction in urban energy emissions. About 480 Gigawatts-thermal of solar water heating capacity was installed globally by the end of

2018 [22]. Research has shown that there are a wide range of domestic hot water (DHW) consumption profiles and these impact the effectiveness of the solar thermal equipment used to produce hot water. The water volume consumed and the heating time is the key variable that impacts domestic solar thermal water heating systems [23-25]. Koroteev and Kharun [26] demonstrated how the dynamic operation of solar thermal collectors can satisfy various DHW consumption profiles. This enables the design of solar collectors to operate with just a fraction of solar coverage.

This technology does not require an urban grid connection infrastructure. It is typically installed as standalone systems using glazed evacuated tubes or flat plate collectors and unglazed systems. These systems are relatively cheap to install and maintain. There are two principal solar hot water system designs employed: open-loop and closed-loop. In open-loop or direct systems, solar radiation is used to heat water directly. In closed-loop or indirect systems, solar energy is initially used to heat transfer fluid, contained within a closed piping loop, which is then subsequently used to heat water in a tank.

3.6   The expectation of artificial photosynthesis to produce useful fuels

One way to potentially generate energy fuels in an environmentally friendly manner is to utilize solar-driven photocatalytic reactions to generate hydrogen or synthetic hydrocarbon biofuels directly by combining water and carbon dioxide feedstocks. These reactions mimic the photosynthesis reactions that sustain the green plant world and are referred to as artificial photosynthesis. Carbon dioxide fixation, in a carbon-neutral system, is conducted together with water oxidation. Consequently, artificial photosynthesis reactions have the potential to remove undesirable carbon dioxide from the atmosphere and produce useful fuel products [27].

However, as yet, there are no commercial-scale artificial photosynthesis system constructed, despite substantial research efforts conducted with a wide range of structures and materials over recent decades [27,28]. This is because the conversion efficiency of artificial photosynthesis reactions remains low. Progress in the miscellaneous fields such as catalysis [6], materials sciences and nanotechnology must be made to improve the efficiency of artificial photosynthesis reactions. Research associated with artificial photosynthesis is ongoing due to the environmental desirability to find cost-effective solutions and technology breakthroughs are likely.

3.7   Evaporative cooling

Cooling spaces in warm climates can be effectively achieved by solar-thermal energy generation, even though solar heat is primarily exploited to produce hot water and space heating. It can also be deployed to support many industrial processes requiring moderately hot or cold conditions and also use solar for drying purposes.

Evaporative cooling is most used in countries that experience seasonal, or year-round, hot humid conditions. It is mainly exploited in the form of air conditioning. It is also exploited in many industrial systems that require cooling towers. Heat can be absorbed by warm air fan-assisted movement over water in the form of latent heat of evaporation of the water. In some systems, evaporative cooling is passive with natural wind energy moving air across open ponds. Cooling towers associated with all thermal power plants (biomass, fossil-fuel and nuclear) all rely on forced or natural evaporative cooling. It is a form of exploitation of solar energy that is usually not recorded in renewable energy statistics; yet, it represents a significant energy requirement.

3.8   Policy and regulatory efforts to promote solar energy uptake

International Energy Agency's Photovoltaic Power Systems Programme (IEA PVPS) (2019) follows and analyzes the policy of photovoltaics annually among its member nations and from time to time releases an updated report addressing the status of government policy associated with this sector (Figure 1) [29].

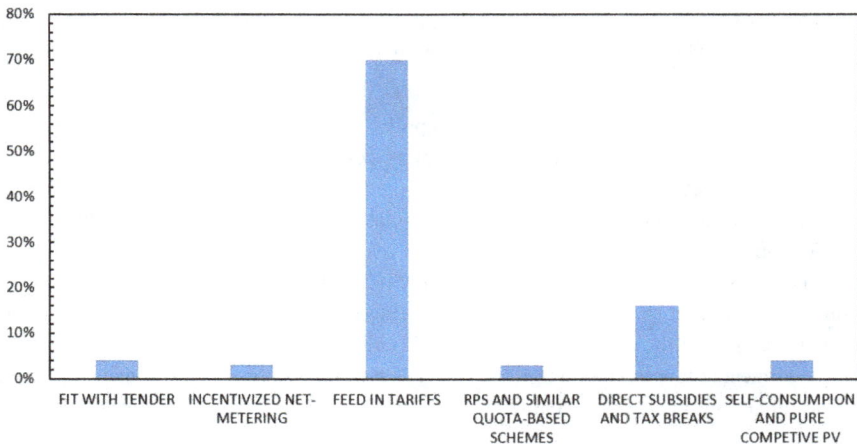

Figure 1 Contribution of market incentives for photovoltaics with the aim of drivers of market share (after Crawley 2013) [29].

Materials for Solar Cell Technologies II   Materials Research Forum LLC
Materials Research Foundations **104** (2021) 134-148   https://doi.org/10.21741/9781644901410-6

Policies employed for photovoltaics tend to be representative of those applying to solar energy more generally [13, 30]. The most developed and utilized government-support mechanism is the feed-in tariff (FiT). The FiT mechanism enables photovoltaic systems to provide energy to a grid or network at a pre-determined price for a specified period [31]. FiTs, therefore, provide PV producers with the guarantee of profitable prices and, thereby, incentivize the construction of more PV capacity.

FiTs have employed extensively in Germany, Italy, United Kingdom, Japan and China. The cost of sustaining them is borne by the consumer, general taxpayer and/or by carbon taxes paid by fossil fuel power producers and utilities.

A renewable portfolio standard (RPS)is a policy backed up by regulation that mandates the increased production and supply of energy from renewable energy sources. When an RPS is in place utilities must honor its requirements or pay penalties for not doing so. An RPS provides a market-based remuneration for renewable electricity producers for their output to the network. Many jurisdictions have allowed projects to remunerate locally produced electricity connected to the grid to take into account local or self-consumption through some form of net-metering mechanisms. These result in a decrease in demand for imported grid electricity and, consequently, a cost-saving for the PV generator/energy consumer and a reduction in cost for the grid utility to supply that component of self-consumed power. Following the successful introduction of RPS, FiT and net-metering in Germany a decade ago, these features are now widely used in many countries around the world.

The way net-metering mechanisms are applied by many utilities often does not adequately compensate PV producers/consumers connected to the grid for the true cost savings they achieve for the grid. For instance, the utilities persist in charging small-scale PV providers high connection charges. Regulators are beginning to recognize the justification to adjust traditional energy-based and capacity-based tariffs to reflect the true value of small-scale PV producers to grids [32]. Nevertheless, the demand for grid-connected solar energy and uptake of net-metering schemes continues to increase due to the enhancement of the energy efficiency of PV cells and systems and the growth of solar energy connected to many power grids [33]. By providing more favorable tariff structures and net-metering adjustments regulators have the potential to motivate higher uptake of small-scale PV installations in many countries. Consequently, many power utilities operating system designed, built and operated to primarily use large-scale fossil-fuel generation are opposed to more favorable net-metering and connection tariff terms for small-scale embedded PV power generators.

Further reductions in the cost of supply of renewable energy generation is desirable. However,the observed continued trends of PV cost reductions are resulting in some difficult decisionsfor policymakers and operational challenges for system operators [34]. For instance, the enhancement of PV power production and the number of intermittent embedded generators tend to increase system management costs, decrease its stability and security of supply. This makes accurate prediction of solar power availability taking into account diurnal and seasonal fluctuations in solar intensity. Sufficient solar capacity and historical solar power generation data exist in Germany that makes it possible to use machine learning algorithms to provide accurate hourly solar power availability forecasts on a regional or country-wide basis [35]. Competition and availability between distributed and centralized power generators become less predictable. As PV costs fall it becomes more difficult for governments and regulators to justify any cost burden on the taxpayers to support new capacity and this can undermine the motivation to achieve the RPS.

Conclusions

Significant progress has been made to address most of the technical challenges facing solar energy technologies, including conversion efficiency, suitable materials and systems configurations, stability, durability and issues related to low and high temperatures. Today, solar energy is exploited in several forms using a range of technologies that continue to improve in terms of their performance due to ongoing innovation and technology refinements. This means that it can be used competitively for various applications. Power generation via photovoltaics, on small and large scales, and concentrated solar thermal for large-scale and sometimes remote applications is a primary use. It is also widely used for water and space heating and cooling on large and small scales. Although energy from the sun is intermittent, it can be successfully stored by a range of thermo-chemical reactions. It can also be usefully transformed by other reactions into valuable fuels and materials but its full potential in this regard has yet to be widely exploited at a commercial scale. Ongoing research into artificial photosynthesis reactions exploiting solar energy to generate clean fuels from a combination of reactions involving water and carbon dioxide requires further technology breakthroughs before it can be established as commercially viable. Integrated policy initiatives and regulatory mandates incorporating feed-in tariffs, net-metering and aspirational renewable portfolio standards have helped to promote and motivate photovoltaic technology uptake in many countries.

References

[1] E. Sadatshojaei, M. Jamial ahmadi, F. Esmaeilzadeh, D.A. Wood, M.H. Ghazanfari, The impacts of silica nanoparticles coupled with low-salinity water on wettability and interfacial tension: Experiments on a carbonate core, J. Disper. Sci. Technol. (2019) 1-15. https://doi.org/10.1080/01932691.2019.1614943

[2] A. Choubineh, H. Ghorbani, D.A. Wood, S.R. Moosavi, E. Khalafi, E. Sadatshojaei, Improved predictions of wellhead choke liquid critical-flow rates: modelling based on hybrid neural network training learningbased optimization, Fuel. 207 (2017) 547-560. https://doi.org/10.1016/j.fuel.2017.06.131

[3] R. Rezaei Dehshibi, A. Sadatshojaie, A. Mohebbi, M. Riazi, A new insight into pore body filling mechanism during waterflooding in a glass micro-model, Chem. Eng. Res. Des. (2019) 151. https://doi.org/10.1016/j.cherd.2019.08.019

[4] E. Sadatshojaei, M. Jamialahmadi, F. Esmaeilzadeh, M.H. Ghazanfari, Effects of low-salinity water coupled with silica nanoparticles on wettability alteration of dolomite at reservoir temperature, Petrol Sci. Technol. 34(2016) 1345-1351. https://doi.org/10.1080/10916466.2016.1204316

[5] D.A. Wood, Microbial improved and enhanced oil recovery (MIEOR): Review of a set of technologies diversifying their applications, advances in geo-energy research. 3 (2019) 122-140. https://doi.org/0.26804/ager.2019.02.02

[6] E. Sadatshojaei, F. Esmaeilzadeh, J. Fathikaljahi, S.E.H. Barzi, D.A. Wood, Regeneration of the midrex reformer catalysts using supercritical carbon dioxide, J. Chem. Eng. 343 (2018) 748-758. https://doi.org/10.1016/j.cej.2018.02.038

[7] M.P. Soriaga, Ultra-high vacuum techniques in the study of single-crystal electrode surfaces, Prog. Surf. Sci. 39 (1992) 325-443. https://doi.org/10.1016/0079-6816(92)90016-B

[8] Chapter 17: Solar energy. Available from: https://personal.ems.psu.edu/~radovic/Chapter17.pdf

[9] J. Kennewell, A. McDonald, The solar constant, 2015, Available from: http://www.sws.bom.gov.au/Category/Educational/The%20Sun%20and%20Solar%20Activity/General%20Info/Solar_Constant.pdf

[10] J. Mankins, Space solar power: The first international assessment of space solar power: Opportunities, issues and potential pathways forward, International Academy of Astronautics. 2011.

[11] M.W. Parker, H. Wynne, N. Beveridge, O. Clint, B. Brackett, S. Gruber, Bernstein energy & power blast: Equal and opposite. if solar wins, who loses?, Bernstein Research. 2014.

[12] Fund announces plans to divest from fossil fuels, Rockefeller Brothers Fund, 2014.

[13]   International energy agency-Photovoltaic power systems programme annual reports for years 2005 to 2018, Snapshot of Global PV Markets. 2019, Available from: http://www.iea-pvps.org

[14] R.M. Swanson, The promise of concentrations, Prog. Photovolt. Res. Appl. 8 (2000) 93-111. https://doi.org/10.1002/(SICI)1099-159X(200001/02)8:13.0.CO;2-S

[15] R. Corkish, W. Lipiński, R. Patterson, Introduction to solar energy, in: G.M. Crawley (Eds.) Solar Energy, Marcus Enterprise LLC, USA (2016) 1-29.

[16] J. Salvatore, World energy perspective: cost of energy technologies, World Energy Council. United Kingdom (2013) 1-48.

[17] V.S. Reddy, S.C. Kaushik, K.R. Ranjan, S.K. Tyagi, State-of-the-art of solar thermal power plants-A review, Renew. Sust. Energ. Rev. 27 (2013) 258-273. https://doi.org/10.1016/j.rser.2013.06.037

[18] P. Pardo, A. Deydier, Z. Anxionnaz-Minvielle, S. Rougé, M. Cabassud, P. Cognet, A review on high temperature thermochemical heat energy storage, Renew. Sust. Energ. Rev. 32 (2014) 591-610. https://doi.org/10.1016/j.rser.2013.12.014

[19] C. Agrafiotis, H. von Storch, M. Roeb, C. Sattler,Solar thermal reforming of methane feedstocks for hydrogen and syngas production-A review.Renew. Sust. Energ. Rev. 29 (2014) 656-682. https://doi.org/10.1016/j.rser.2013.08.050

[20] S. Sadhishkumar, T. Balusamy, Performance improvement in solar water heating systems-A review, Renew. Sust. Energ. Rev. 37 (2014) 191-198. https://doi.org/10.1016/j.rser.2014.04.072

[21]L. Pérez-Lombard, J. Ortiz, C. Pout, A review on buildings energy consumption information, Energ. Build. 40 (2008) 394-398. https://doi.org/10.1016/j.enbuild.2007.03.007

[22]   REN21 Renewables 2019 global status report. REN21 Secretariat, Paris, (2019).

[23]   A. Henning, Equal couples in equal houses: Cultural perspectives on Swedish solar and bio-pellet heating design, in: S. Guy, S. Moore (Eds.), Sustainable Architectures: Cultures and Natures in Europe and North America,Spon Press, Routledge. (2005) 103-118.

[24]   R. Spur, D. Fiala, D. Nevrala, D. Probert, Influence of the domestic hot-water daily draw-off profile on the performance of a hot-water store, Appl. Energ. 83 (2006) 749-773. https://doi.org/10.1016/j.apenergy.2005.07.001

[25] J. Burch, C. Christensen, Towards development of an algorithm for mains water temperature, in Proceedings of the solar conference. Citeseer, (2007).

[26] D. Koroteev, M. Kharun, Influence of construction of transparent covering on efficiency of concrete heat treatment in shuttering forms with using solar energy, Structural mechanics of engineering constructions and buildings. 14 (2018) 64-69. https://doi.org/10.22363/1815-5235-2018-14-1-64-69

[27] W. Tu, Y. Zhou, Z. Zou, An in situ simultaneous reduction-hydrolysis technique for fabrication of TiO_2-graphene 2D sandwich-like hybrid nanosheets:graphene-promoted selectivity of photocatalytic-driven hydrogenation and coupling of CO_2 into methane and ethane, Adv. Mater. 26 (4607) (2013). https://doi.org/10.1002/adfm.201202349

[28] Z. Han, R. Eisenberg, Fuel from water: The photochemical generation of hydrogen from water. Acc. Chem Res. 47 (2014) 2537-2544. https://doi.org/10.1021/ar5001605

[29] G.M.Crawley, The world scientific handbook of energy, 3, World Scientific Publishing, Hackensack (2013).

[30] M. Miller, S. Cox, Overview of variable renewable energy regulatory issues, National Renewable Energy Laboratory (2014).

[31] M. Mendonça, Feed-in Tariffs, Accelerating the deployment of renewable energy, Sterling, VA: Earthscan, (2007).

[32] R. Passey, M. Watt, R. Brazzale, Impacts of PV, AC, Other technologies and tariffs on consumer costs, Australian PV Institute (APVI), (2013).

[33] J.P. Marshall, Disordering fantasies of coal and technology: Carbon capture and storage in Australia, Energ. Pol. 99 (2016) 288-298. https://doi.org/10.1016/j.enpol.2016.05.044

[34] S. Vinci, D. Nagpal, R. Ferroukhi, E. Zindler, A. Czajkowska, Adapting renewable energy policies to dynamic market conditions, International renewable energy agency (IRENA) (2014).

[35] D.A. Wood, German solar power generation data mining and prediction with transparent open box learning network integrating weather, environmental and market variables, Energ. Convers.Manag. 196 (2019) 354-369. https://doi.org/10.1016/j.enconman.2019.05.114

Materials for Solar Cell Technologies II
Materials Research Foundations104 (2021) 149-174

Materials Research Forum LLC
https://doi.org/10.21741/9781644901410-7

Chapter 7

Hybrid Materials for Solar Cells

Umesh Fegade[1*], Ganesh Jethave[2]

[1]Bhusawal Arts, Science and P. O. Nahata Commerce College Bhusawal, MH, India 425201

[2]School of Environmental and Earth Sciences Kavayitri Bahinabai North Maharashtra University Jalgaon, MH, India 425001

* umeshfegade@gmail.com

Abstract

Solar energy is an attractive renewable energy source across the globe that can help overcome the energy crises and has the ability to replace conventional resources. Hybrid solar cells have higher conversion efficiency. In the current chapter the research related to the carbon nanotubes, organic and inorganic solar cell, dye-sensitized solar cells and tandem solar cells are reviewed. The organic solar cells are most suitable and economic, but it has low efficiency of up to 15%. The inorganic solar cells are very expensive and have high efficiency of up to 46% and are used in space applications. The hybrid solar cell is the third type and the perovskite tandem has already proven to be quite efficient (17%) and low cost, mostly because of the cheap materials that are being used.

Keywords

Renewable Energy, Hybrid Solar Cell, Carbon Nanotube, Organic and Inorganic Solar Cell, Dye Sensitized Solar Cell, Tandem Solar Cell

Contents

1.   Introduction

Cost-effective and environmentally sustainable renewable energy generation remains a major challenge for both technological and scientific development [1-3]. Solar energy will be a significant contributor to future energy production[4,5].Solar power is clean, renewable and sustainable since no carbon dioxide is produced [6]. In order to be a huge energy source to compete with fossil fuels, the cost of producing solar cells must still be reduced. The effort to create new materials for solar cells has been difficult, and recently new classes of photovoltaic materials have been discovered. The hybrid solar cell is one of the latest most commonly produced photovoltaic(PV) cells [5]. The introduction of a combination of polymers and nanoparticles in cell designs is rapidly replacing silicon PV modules to increase efficiency and reduce the cost of PV systems[7].

PV cell development is traditionally split into 3 generations. The first generation is the era of crystalline silicon cells. Second is the thin-film solar cells generation and in the third PV generation, concentrating photovoltaic (CPV), dye-sensitized solar cells (DSSC), organic-inorganic hybrid solar cells, perovskites etc. are beginning and has yet to be commercialized to an extent [8]. CPV has the capability of having the highest performance of any PV module, though it is not clear at what value. Other organic or hybrid organic/traditional (DSSC) PV technologies are at the research and development (R&D) stage [9]. These offer lower performance, but also lower costs and weights, and offer free-form sizes; thus could fill the market [10].

The attributes of organic as well as inorganic semiconductors are incorporated into hybrid solar cells. Photovoltaics are hybrids with organic material consisting of conjugated polymers which absorb the light and act as an electron donor. The acceptor and transporter in the hybrid solar cells are the inorganic materials [11,12]. Innovation is being shown in the use of carbon nanotubes (CNTs) in PV systems. Because of their unique tubular structures and dimensions, carbon nanotubes exhibit attractive mechanical and electronic properties[13].Transparent conduction electrodes made from CNTs are a fine substitute for indium-tin-oxide electrodes. Charges in the carbon nanotubes help to improve the carrier's mobility and conductivity so that the recombining of photo-generated charge improve the photovoltaic effect[14,15].

The dye-sensitive solar cells as high energy performance and low-cost material have been taken into consideration in recent years[4,16]. The research activity on solar cells, organic-inorganic photovoltaic devices has increased exponentially, especially since the discovery of the dye-sensitive nanocrystalline solar cell. A new record-breaking conversion performance of 11 % has already been reached, but its practicality remains difficult in the electrolyte instability[15,17–19].

Tandem solar cells use a large bandgap material as a front cell and a small bandgap material as a rear cell to minimize excess thermal energy [20,21]. In this chapter, we have discussed and reviewed hybrid material for solar cells consisting of carbon nanotubes, DSSC, hybrid organic–inorganic solar cells, polymer-nanoparticle composite-based solar cells, etc.

2.   Types of hybrid cell based on material used in it

2.1   Carbon nanotubes (CNTs)

Metal decorated carbon nanotubes have multi uses such as exciton carrier within a polymer-based photovoltaic layer and semiconducting CNT as the photoactive layer[14,22–24]. The CNTs in donor and acceptor form photovoltaic systems have been developed with functionalized multiwalled carbon nanotubes by Pradhan and co-workers [25], and also developed the heterostructure devices between composite polymer CNT and buckminsterfullerene or preferably C_{60} layers. CNTs in the CNT/C_{60} polymer framework produced a greater volume of exciton dissociation and increased mobility for carrier transport [25].

Somani and co-workers [26] presented Platinum nanoparticles in organic/organic-inorganic n-Si/poly (3-octylthiophene) (P3OT/n-Si) hybrid solar cell (HSC) in a photovoltaic application (Figure 1) [26]. They found that a multi-wall carbon nanotube with platinum (MWCNT-Pt) is a better contender than pristine MWCNT for inclusion in P3OT/n-Si heterojunction solar cells to improve their efficiency. Conversion efficiency with MWCNT-Pt = 0.775% while in MWCN conversion efficiency = 0.145% [27].

By using a Pt nanoparticles-adsorbed carbon nanotube yarn as a counter electrode, Zhang et al. [28] have obtained 4.85% power conversion efficiency with normal illumination (AM1.5, 100 mW / cm^2) (Figure 2) [28]. According to the authors platinum (Pt) nanoparticles adsorbing in a porous nanotube yarn results in the increased interfacial region of Pt/electrolyte and significantly reduces the resistance of charging movement across the electrolyte interface. It was proposed that such yarns could minimize the use of noble metals, reduce the weight of the equipment and improve the efficiency of the solar cells.

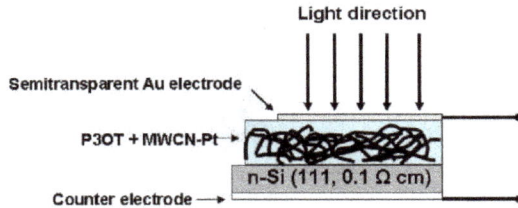

Figure 1 P3OT/n-Si solar cell (Reprinted with the reference Somani et. al., 2008, permission of AIP Publishing)

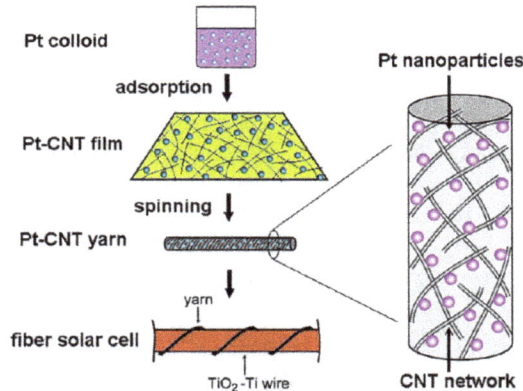

Figure 2 Illustration of the preparation processes of Pt-adsorbed hybrid yarns and application as counter electrodes for fibre solar cells. (Reprinted from the reference [28], Copyright ACS)

MWCNT hybrid is synthesized by Arbab et al. [29]as an activated carbon (AC)-doped composite electrode (CE) for quasi-solid state solar cells with dye sensitization (DSSCs) (Figure 3) [29]. The MWCNTs model, with AC-doped, displayed a considerable improvement in electrocatalytic (ECA) activities towards electrolyte polymer gel, which demonstrated a strong mechanism for electron transportation. An inappropriate amount of quasi-solid electrolyte may be caused by inadequate oxygen surface groups and defect rich structure, and multiple locations to catalytic iodide/triiodide reactions. The corresponding DSSC compound of this counter electrode was 10.05% effective with a large fill factor (83%)(Figure 4)[29].

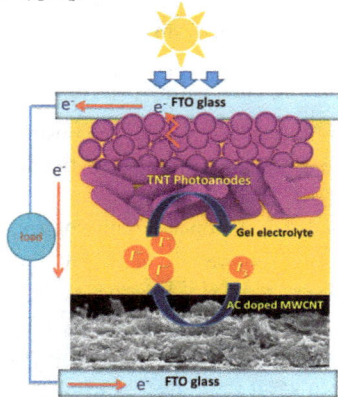

Figure 3 AC-doped MWCNT based DSSCs. (Reprinted from the reference [29], Copyright ACS)

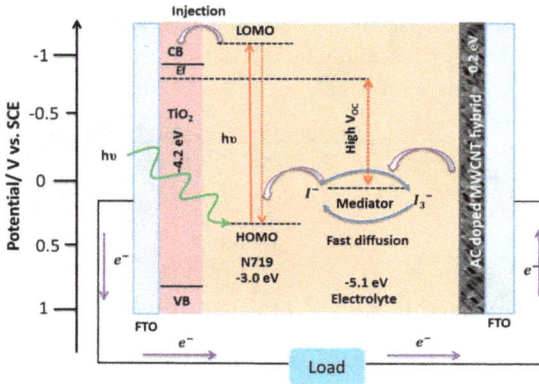

Figure 4Photocurrent generation in DSSCs. (Reprinted from the reference [29], Copyright ACS).

2.2　Organic-inorganic solar cell

The hybrid solar cell is considered to be a viable alternative to low-cost photovoltaic devices, as Schottky junctions could be formed in an organic and inorganic matter [30]. The production of hybrid solar cell structures has been carried out through inorganic materials including quantum dots, nanoparticles and nanowires [31–36].

High-strength conversion efficiency has been demonstrated by Yodying Yong et al. [37]in inorganic-organic solar cells encapsulating titanium dioxide (TiO_2)filled with poly(3-hexylthiophene) (P3HT) and [6,6]-phenyl C61-butyric acid methyl ester (P3HT/PCBM) as the active layer. An open-circuit voltage, short-circuit current density, fill factor and power conversion efficiency of the high-grid solar cell, respectively, is 646mV, $9.95mAcm^{-2}$, 51.6% and 3.32%[37].

Balis and coworkers[38]made a cell by combining nc-TiO_2, P3HT and cadmium sulfide (CdS) quantum dots, in the presence of 2-amino-1-methylbenzimidazole (AMBIm) with an efficiency that reached 0.87±0.13%. Cells exhibited amazing stability and power conversion efficiency [38].

Jayan and Manthiram [39]have shown the performance and stability of P3HT-TiO_2-based cells (Figure 5) [39] can be improved by using copper (Cu) as the top hole collecting electrode in the absence of oxygen (Figure 6) [39]. These results can contribute to the synergy of hybrid solar cell stability. This research nevertheless reflects a preliminary study of the Cu effect on the stability of hybrid solar cells with P3HT-TiO_2 in an inert atmosphere [39].

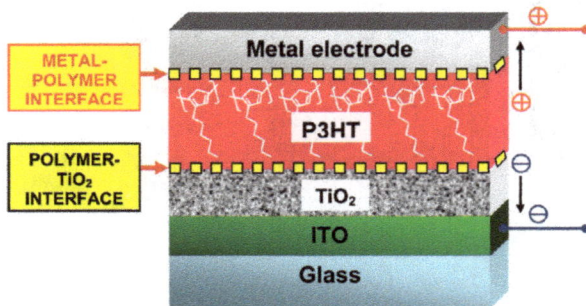

Figure 5 Interfacial prototype of P3HT-TiO₂ solar cell. (Reprinted with permission from reference [39], 2011 Copyright ACS).

Materials Research Forum LLC
https://doi.org/10.21741/9781644901410-7

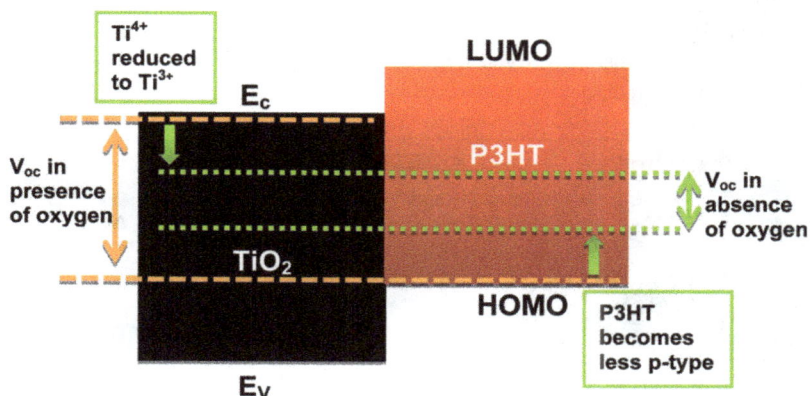

Figure 6 Energy band diagram of P3HT-TiO₂. (Reprinted with permission from reference [39], 2011 Copyright ACS).

Sequential hybridized polymer solar cells are fabricated by Liao and co-workers[40](2012) using a rutile TiO_2 nanoarchitecture array. The diotic molecule plays an important role in the interface morphology of the TiO_2 nanorod (NR)/P3HT hybrid, becoming a photocurrent contributor. *cis*-Ru(H_2dcbpy)(dnbpy)(NCS)₂(Z907) and Indoline dye (D149) dye molecules not only provide proper band alignment between P3HT/TiO₂NR, but also improve the consistency of the hybridized interface morphology. Three-dimensional TiO_2 nanodendrite (ND) arrays have been further assigned to expand the interface area of the planned heterojunction hybrid solar cell. A remarkable efficiency of 3.12% is found in the D149-modified TiO_2 ND array/P3HT hybrid solar cell [40].

Almohsin and Cui [41] investigated a graphene-enriched P3HT (G-P3HT) and porphyrin grafting of the zinc oxide (ZnO) NW hybrid structure for applications in the solar cell which significantly improves the cell efficiency. According to their study, the cell efficiency increases from 0.09 to 0.4%. Using modified porphyrin which is responsible for improved charge injection and increased light absorption at the junction interface(Figure 7) [41]. This study shows that hybrid structures have promising applications for fabricating high-efficiency solar cells [41].

Greaney and coworkers [42] fabricated P3HT:CdSe-(tBT) hybrid solar cells devices revealed efficiencies of 1.9% and a maximum open-circuit voltage (V_{OC})=0.80 V. The open-circuit potential associated with devices using the cadmium selenide (CdSe) tert-Butylthiol (tBT) acceptors is recognized by lowest unoccupied molecular orbital

(LUMO) and highest occupied molecular orbital (HOMO) energy of P3HT. The authors showed a direct relationship between the LUMO energy displacement induced by the ligand and the open circuit potential observed in functional electrical cells [42].

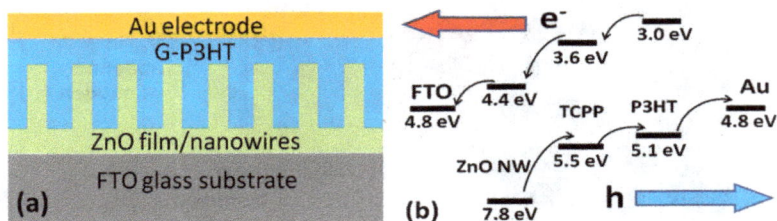

Figure 7Schematic diagram (a) energy charge transfer (b). (Adapted from reference [41], 2012. Copyright ACS).

Jeong et al. [43] revealed the Si nanocones and poly (3, 4-ethylenedioxythiophene): polystyrenesulfonate(PEDOT:PSS) hybrid solar cell as conductive polymer (Figure 8)[43]. Nanocones arrangement with an aspect ratio of less than two tailored structures for enhancing the absorption of light, and providing excellent antireflection and light diffusion results. The efficiency of power conversion above 11%, the optimum nanoconic structures for the thickness of 10 μm, if the Solar Cell has a short-circuit current density of up to 39.1 mA/cm^2 (Figure 8b). Hybrid Si nanocone/PEDOT: PSS solar cells are promising as a cost-effective alternative solution to the process with a very thin material [43].

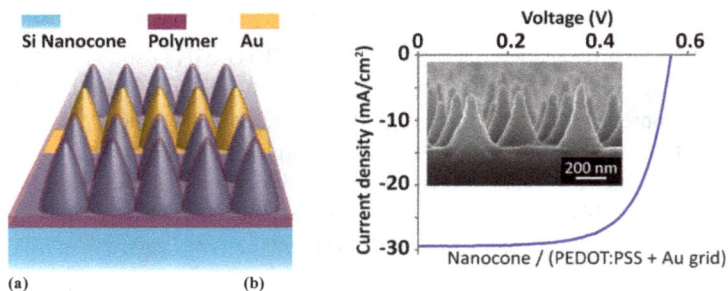

Figure 8(a) Fabricated Si nanocone/polymer solar cell. (b) Characteristics of Si nanocones coated with PEDOT: PSS and Au grid. (Adapted from reference [43], 2012 Copyright ACS).

Pudasaini et al. [30] described hybrid solar cell device SiNP/PEDOT: PSS. power conversion efficiency (PCE)was found to be 9.65%. With the use of the ultrathin atomic layer deposition (ALD) deposited aluminum oxide (Al_2O_3) junction activation layer (Figure 9) [30], the short-circuit current density, open-circuit voltage, PCE value and fill factor were observed to be 30.1mA/cm^2 and 578mV, 10.56%, 71% respectively [30].

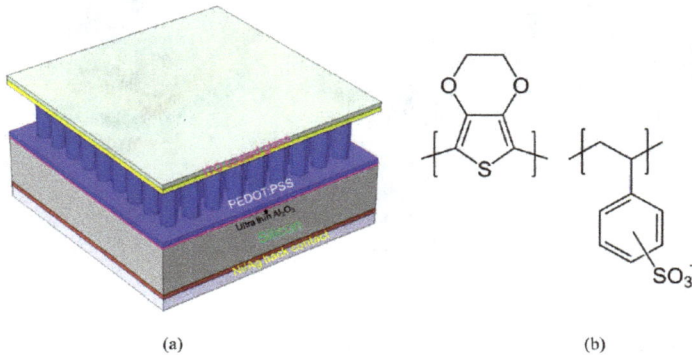

(a)   (b)

Figure 9 (a) Schematic illustration of the fabricated SiNP/PEDOT: PSS solar cell. (b) Chemical structure of PEDOT: PSS. (Reprinted with permission from reference [30], 2013 Copyright ACS).

Eom et al. [44] first, inverted bilayer HSC of ZnO:P3HT IMs, ZnO act as n-type semiconductor and in second, inverted BHJ solar cell of ZnO/P3HT:PCBM, it roles as an electron transport layer/hole blocking layer(ETL/HBL). After the interfacial modification in the ZnO:P3HT bilayer device (Figure 10) [44], and ZnO/P3HT:PCBM based bulk heterojunction (BHJ) device, with the efficiency of 0.42% and 4.69% is achieved, respectively. The authors found that IM's help to reduce recharge and leakage by reducing the number of defect locations and trap and improving hydrophilic ZnO [44].

Qin et al. [45] reported graphene quantum dots with bright blue emission and showed a slight difference in the carrier band between TiO_2 and P3HT. This bulk heterojunction of high-energy photons increased the peak by incorporating graphene quantum dots (GQDs) into the hybrid heterojunction as an additional buffer layer. This intense mix gave 3.16% more PCE, than traditional TiO_2/P3HT devices. GQDs unique band structure funnels a photoexcited charge carrier directly into the circuit. Although this function is far from optimized, it offers profound advantages, as well as the relative ease of electronic structures, low toxicity, and promising candidate buffer material for scalable photovoltaic applications [45].

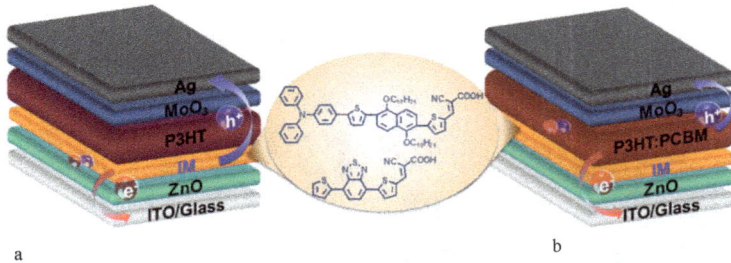

a   b

Figure 10 The basic device structures of hybrid solar cells in this study: (a) ZnO:P3HT bilayer structure and (b) P3HT:PCBM BHJ structure with ZnO as ETL/HBL. (adapted from reference [44], 2014, permission ACS).

Lee and co-workers [46] demonstrated a high-performance Si-organic hybrid heterojunction solar cell utilizing TiO_2 interlayer between PEDOT:PSS and Si nanoholes to produce a conformal contact on the surface of the Si nanostructure. The high Voc=0.63V was attributed to the surface passivation of silicon (Si) by annealed TiO_2, with this value of Jsc=35.7mA/cm^2 and PCE=14.7% were obtained. Authors convinced that with proven surface passivation and conformal PEDOT: PSS/Si nanohole interfaces for enhanced contact, this Si-organic hybrid heterojunction solar cell with solution-processed TiO_2 interlayers has excellent potential for application as a high-efficiency and low-cost Si solar cell [46].

Um and co-researchers [47] demonstrated an sliver/silicon dioxide (Ag/SiO_2) electrode embedded Si substrate cells (Figure 11) [47]. With a 607mV open-circuit voltage and a 34.0 mA / cm^2 short-circuit current density; the 1 cm^2 solar hybrid cell has an output of up to 16.1%. In particular, the fill factor of this solar cell rose significantly by 78.3 %, compared to the fill factor of traditional electrodes (61.4 %), because the metal/contact resistance of the 1μm-thick Ag electrode was dramatically reduced [47].

Shin and co-workers [48] developed a multilayer TCE/PEDOT: PSS/Si nanowires/n-Si/TiOx hybrid solar cells (HSCs) to resolve problems (deficiencies) of HSCs for practical applications. Highest power conversion efficiency (PCE)10.14% was achieved without TiO_x at layer number 2 (Figure 12) [48]. Remarkable inhibition of the carrier recombination enhancing PCE to 12.10% because of the Si–O–Ti chemical bond formed by deposited TiO_x layer on the backside of the Si. PCE also enhanced by enhancing

Jscdue to the use of silicon nanowire. Resulting PCE is mainly due to the enhancing Jsc and inhibition of recombination [48].

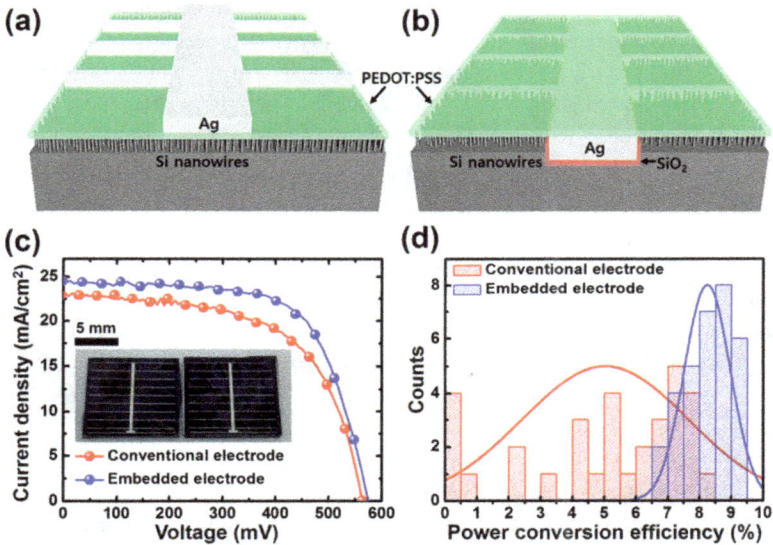

Figure 11 PEDOT:PSS/Si hybrid cells (a) conventional (b) metal electrodes. (c) Curves (d) efficiency distributions. (adapted from reference [47], 2017. permission ACS).

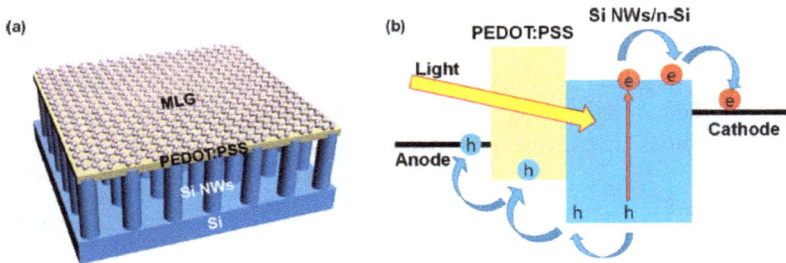

Figure 12 (a,b) Schematic and energy band diagram of a typical MLG/PEDOT: PSS/Si NWs/n-Si/InGa HSC. (Reprinted with permission from reference [48], 2018, Copyright ACS).

Materials for Solar Cell Technologies II
Materials Research Foundations104 (2021) 149-174

Materials Research Forum LLC
https://doi.org/10.21741/9781644901410-7

2.3   Dye sensitized solar cells

Dye-sensitized solar cells (DSSC) play the leading role in photovoltaic system research with advantages such as the requirement of largely plentiful earth materials, respectable efficiencies, low demand for high-temperature material processing and cell assembly simplicity. The electricity generation process and cell assembly procedures have therefore, been tested well in a typical DSC. They have extremely economical applications.

Brown et al. [49] have identified a successful strategy for combining metallic nanoparticles with a strong resonance of surface plasmon in solar cells sensed by the dye that exceeds four major numbers of metal insertion in most solar cells: recombination within the metal, thermal stability during processing, chemical stability and control of metallic nanoparticles, that separates the chromosomes from the dye to discourage non-radioactive cooling. According to the authors, this new plasmonic photoelectric system led to better absorption of light and the generation of light current in solar cells that are dyed-sensed in nature [49].

Wang et al. [50]fabricated etching patterned (230nm of etching depth) at nano level on commercial fluorine-doped tin oxide (FTO) by nanoimprint lithography and reactive ion etching (RIE) to obtained nanopillar and nanoline arrays which increase FTO resistance (Figure 13) [50]. Authors observed enhanced light scattering from nano-patterned FTO in the wavelength ranges from 400−650nm and improved power conversion efficiency about 2−5% of DSSCs, due to light scattering resulted across the interface with 4μm TiO_2 layer in the same wavelength range[50].

Triangular silver nanoprisms were used by Gangishetty and co-workers [51] for light harvesting efficiency (LHE) in a DSSC (Figure 14) [51]. Authors concluded that overall PCE (32 ± 17%) of the device improved by loading of 0.05% Ag@SiO_2 in the titania photoanode. Such variations were further enhanced at the advance wavelength in LHE according to the authors and further increased Ag@SiO_2 loading, which would have a detrimental effect on the performance of electron collection [51].

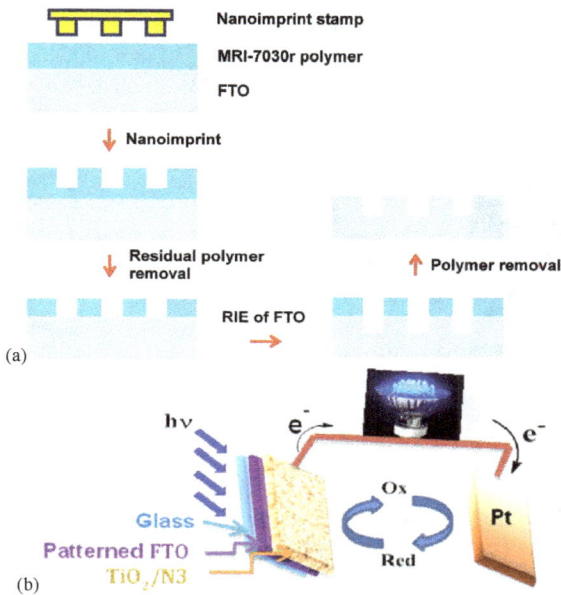

Figure 13 (a) FTO patterning procedure using NIL and RIE. (b) Schematic illustration of mechanism of the cell (Reprinted with permission from reference [50] 2012 Copyright ACS).

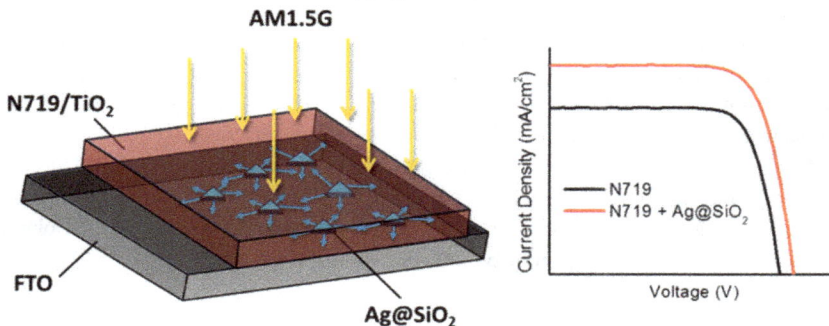

Figure 14 (a) Schematic illustration of fabrication of the cell (b) Current−voltage (J−V) characteristics of the highest performing DSSC. (Reprinted with permission from reference [51] 2013, Copyright ACS).

Materials for Solar Cell Technologies II   Materials Research Forum LLC
Materials Research Foundations **104** (2021) 149-174   https://doi.org/10.21741/9781644901410-7

Kelkar et al. [52] utilized chemically modified egg protein albumen in the quasi-solid state DSSC through appropriate molecular engineering of cross-linking property (Figure 15) [52]. Authors claimed a significantly higher 5.75% efficiency of functionalized egg albumen based gel DSSC than a non-functionalized DSSC having 4.6% efficiency [52].

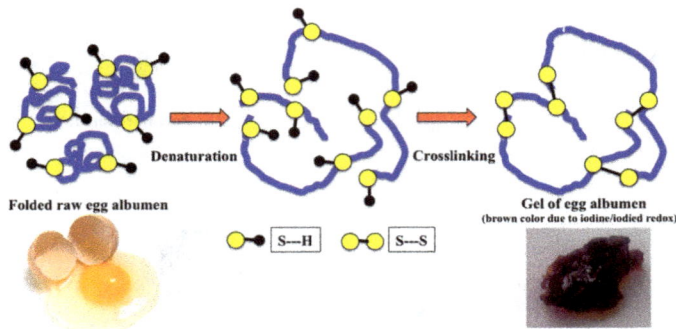

*Figure 15 Schematic showing denaturation and cross-linking of egg albumen.
(Adapted from reference [52], 2014, Copyright ACS).*

Karam and co-workers [53] used urchins-like ZnO nanowires (NWs) in the DSSC as photo-anode, in the form of multi-layer, with TiO_2 nanostructures, and core-shell nanostructures have been sensitized to the Ru dye (N719) solution. The photon absorption improves with the number of layers, according to the scientists, but the photovoltaic capacity declines as a consequence of the recombination of the charge. The monolayer of urchins thus demonstrated the maximum efficiency; this can probably be due to the low electric connection between urchin layers. The low conductivity of these nanostructures in the core-shell is emphasized. The authors opined that further optimization can improve the overall efficiency of photoconversion. Similarly Kang et al. [54] prepared mesoscopic nickel(II) oxide (NiO) films and employed them as photocathodes in p-type DSCs. Authors observed good results than others and claimed that this is the first anodic NiO film synthesized with mesoscopic architecture [53].

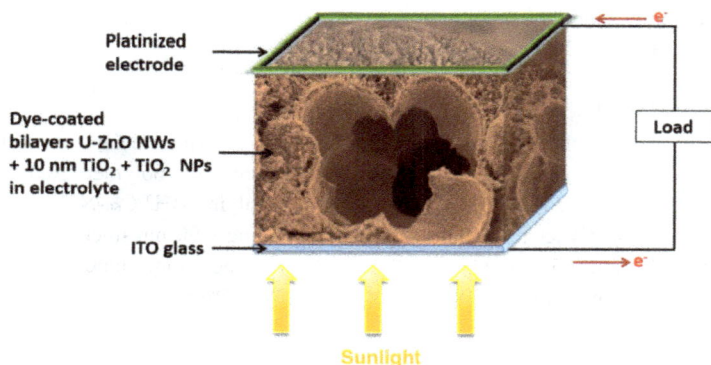

Figure 16 Schematic illustration of fabrication and mechanism of the cell. (Reprinted with permission from reference [53], 2018, Copyright ACS).

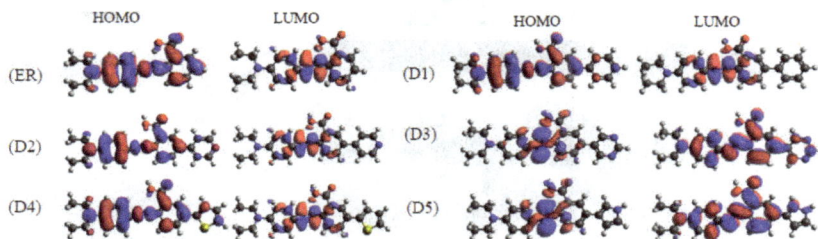

Figure 17 HOMO and LUMO of the dyes. (Reprinted with permission from reference [55], 2019, Copyright (2019) Elsevier)

Computational study of AL-Temimei and Alkhayatt [55] has reported series of dye derivative from the ethyl red dye have been studied. The proposed dyes show applicable energy, photoelectric and spectroscopic feature, since new dyes' lowest unoccupied molecular orbital (LUMO) levels are much higher than TiO_2 conduction band (CB), where as the highest occupied molecular orbital (HOMO) energy levels are lower than the I^-/I_3^- systems redox potential (Figure 17) [55]. This means that the recharge of the colour after the photo-oxidation reaction and the ease of transmitting the electron from the donor to the receiver is easy and that regeneration is possible after this. The investigators found that the dyes investigated were beneficial for the development of new materials for organic solar cells, with greater photoelectrical strength and a strong photovoltaic output[55].

2.4 Tandem solar cells

Perovskite-based tandem solar cells draw attention to their performance capacity. Liu et al. [56] achieved a maximum efficiency of 16.0% perovskite/polymer solar cell (ITO/PEDOT:PSS/Perovskite/PC61BM/C_{60}-SB/Ag/MoO_3/PolymerBHJ/C_{60}-N/Ag) (Figure 18) [56] is 75% higher than that of the corresponding ~90 nm thick perovskite single-junction device and 65% higher than that of the polymer single junction device. Authors achieved a very high V_{OC} of 1.80 V and a high FF of 77% [56].

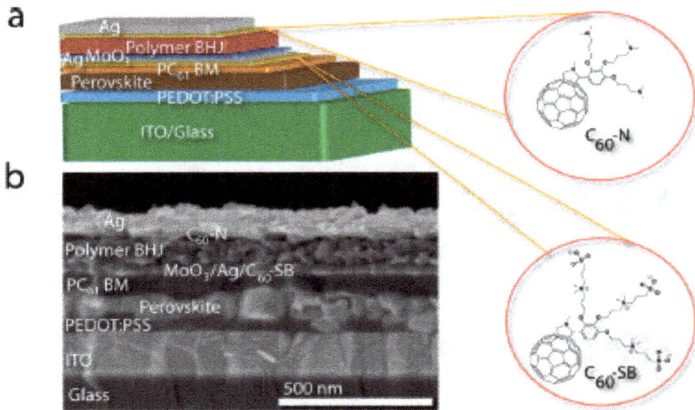

Figure 18 (a) Fullerene interlayer materials (C_{60}-N, C_{60}-SB). (b) False color cross-sectional SEM image. (Adapted from reference [56] 2016, Copyright ACS).

Kranz et al. [57] produced a close infrared transparent perovskite with a wavelength of 800 to 1000 nm. The authors stated that a tandem unit with 19.5 % efficiency is feasible, owing to the combination of the near-infrared transparent perovskite top with the CIGS (Cu(In, Ga)Se_2), bottom layer (Figure 19) [57]. There are suggestions and possibilities for systems with capability approaching 27 % of the future development of perovskite/CIGS tandem systems [57].

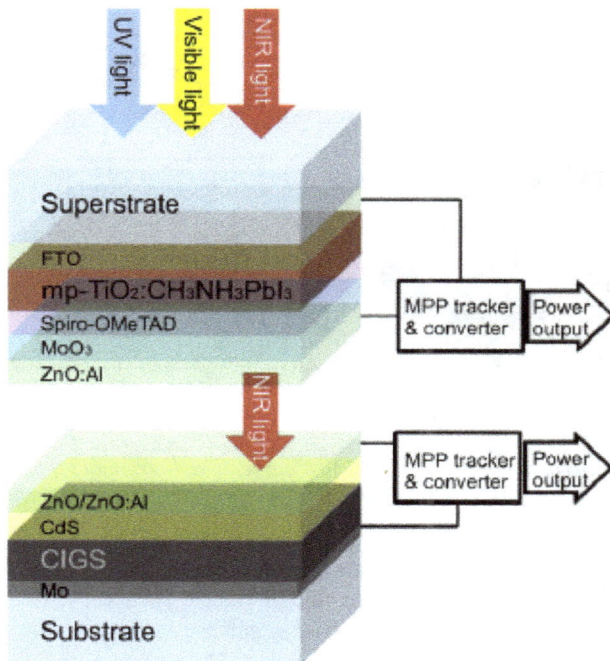

Figure 19 Schematic of a stacked 4-terminal perovskite/CIGS tandem solar cell. The top cell is a perovskite cell in superstrate configuration and the bottom cell is a CIGS cell in substrate configuration. (Reprinted with permission from reference [57], 2015, Copyright ACS)

Guchhait et al. [58] demonstrated 16 % efficiency of the semi-transparent perovskite solar cell, plus visible light range transparency by 12%, and the near-infrared region transparency by more than 50%. The authors used an optimized perovskite cell for a 4 T (4-terminal) tandem perovskite / Cu-In-Ga-Se$_2$(CIGS) solar cell (Figure 20) [58] with efficiency (un-certified) of 20.7%, as against a single CIGS cell with an absolute efficiency improvement of more than 7 % [58].

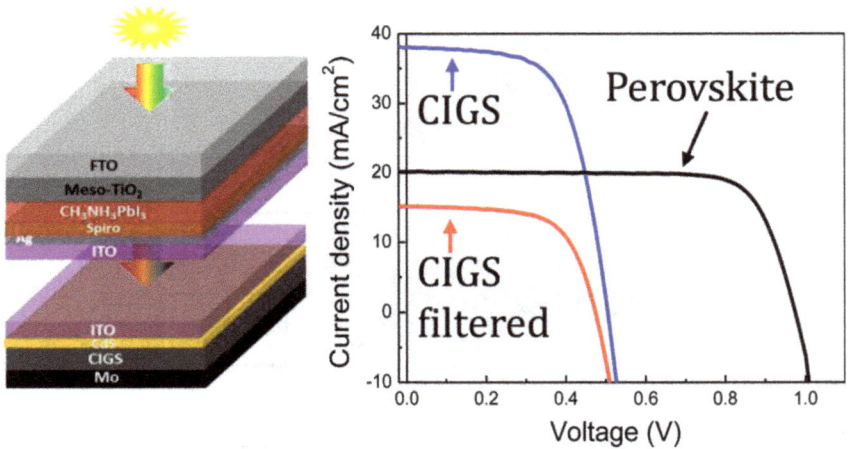

Figure 20Tandem schematic and J−V curve, of a 4-terminal tandem perovskite/CIGS combination. (Adapted from reference [58], 2017, Copyright ACS).

Introduction of the nano-pyramid structure between the top and the bottom cells, Li and coworkers [59] proposed ultrathin a-Si/c-Si tandem solar cells with an effective light trapping design. According to the authors, significant improvement in the current density of short circuits for the top (48%) and bottom (35%) was observed as a result of superior cell light collection. Of ultrathin tandem cells with a minimum of 8 μm of silicon, transfer efficiencies of up to 13.3 %, this is 29 % better than for a planar cell [59].

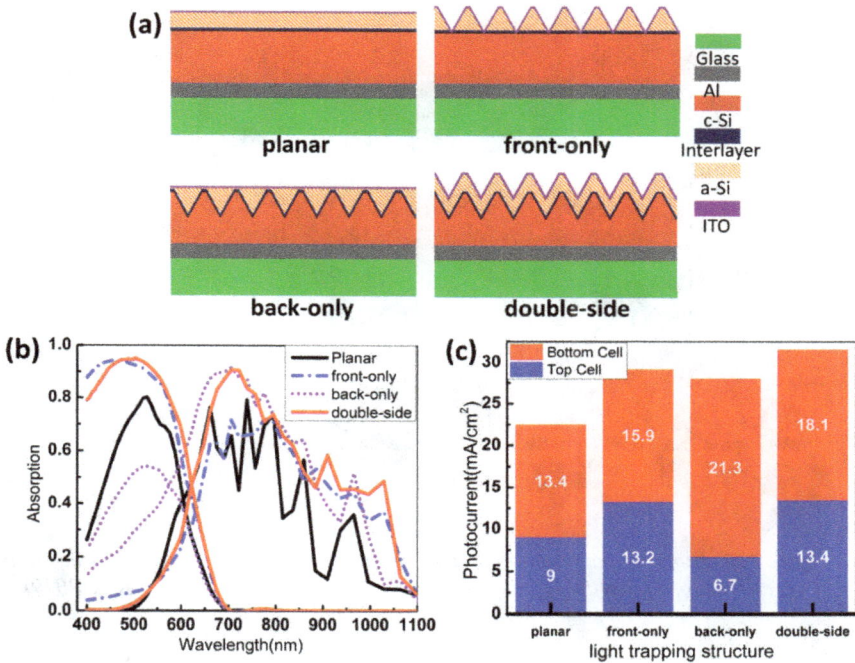

Figure 21 (a) Four tandem cell structures: planar; (b) absorption spectra for the cells; (c) photocurrent density for different cells. (Reprinted with permission from reference [59], 2014, Copyright (2014) American Chemical Society)

The output of a hydride vapor phase epitaxy (HVPE)-grown GaAs top cell embedded in a 4-terminal GaAs/Si tandem system, with an efficiency of 29 %, was evaluated by Vasant [60]. By comparing the results with the tandem cell of metalorganic vapor phase epitaxy (MOVPE)-grown GaAs/Si, the authors improved the HVPE-grown GaAs model and give a short-term utility with low-cost III-V deposition techniques for the 31.4 % range [60].

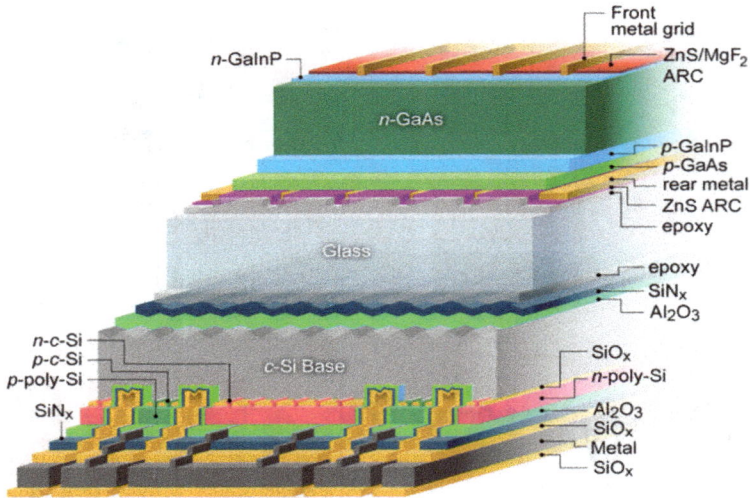

Figure 22 Schematic cross-section of a 4T GaAs//Si tandem device with Al-free GaAs top cell. (Reprinted with permission from reference[60], 2019, Copyright (2019) American Chemical Society)

Conclusion

The researchers attempt to utilize the full range of sunlight for conversion into electricity. The use of multilayer of dyes increases the broad solar spectrum absorption on one hand, however, an increase of layer thickness significantly restricts the solar cell performance. For large-scale development, improvement in the output of hybrid cells is a must. Efficiency can be improved by optimizing several variables. The reduction of the bandgap in order to absorb infrared photons that contain a considerable fraction of the solar spectrum energy must be explored. 70% quantum efficiency for the blue photons has been demonstrated by organic photovoltaic.

References

[1] S.B. Darling, F. You, T. Veselka, A. Velosa, Assumptions and the levelized cost of energy for photovoltaics, Energ. Environ. Sci. 4 (2011) 3133–3139. https://doi.org/10.1039/c0ee00698j

[2]   J. Fan, B. Jia, M. Gu, Perovskite based low-cost and high efficiency hybrid halide solar cells, Photonics Res. 2 (2014) 111. https://doi.org/10.1364/prj.2.000111

[3]   D. Yue, P. Khatav, F. You, S.B. Darling, Deciphering the uncertainties in life cycle energy and environmental analysis of organic photovoltaics, Energ. Environ. Sci. 2 (2012) 9163–9172. https://doi.org/10.1039/c2ee22597b

[4]   C. Chen, Y. Lu, E.S. Kong, Y. Zhang, S.T. Lee, Nanowelded carbon-nanotube-based solar microcells, Small. 4 (2008) 1313–1318. https://doi.org/10.1002/smll.200701309

[5]   B.W. Park, B. Philippe, X. Zhang, H. Rensmo, G. Boschloo, E.M.J. Johansson, Bismuth based hybrid perovskites $A3Bi_2I_9$ (A: Methylammonium or Cesium) for solar cell application, Adv. Mater. 27 (2015) 6806–6813. https://doi.org/10.1002/adma.201501978

[6]   Giorgio Dodero, IPGSRL " 2011 India energy handbook",(2011).

[7]   S. Sawant, The properties and advantages of the hybrid solar cell, Int. Res. J. Sci. Eng. A5 (2018) 19–25.

[8]   E. Da Como, F. De Angelis, H. Snaith, A. Walker, unconventional thin film photovoltaics, RSC(2016). https://doi.org/10.1039/9781782624066-fp001

[9]   S.K. Balasingam, M. Lee, M.G. Kang, Y. Jun, Improvement of dye-sensitized solar cells toward the broader light harvesting of the solar spectrum, Chem. Commun. 49 (2013) 1471–1487. https://doi.org/10.1039/c2cc37616d

[10] L. El Chaar, L.A. Lamont, N. El Zein, Review of photovoltaic technologies, Renew. Sustain. Energy Rev. 15 (2011) 2165–2175. https://doi.org/10.1016/j.rser.2011.01.004

[11] M.T. Lloyd, Hybrid solar cells, in: Encycl. Nanotechnol., Springer Netherlands, 2016, pp. 1494–1500. https://doi.org/10.1007/978-94-017-9780-1_14

[12] S.E. Shaheen, D.S. Ginley, G.E. Jabbour, Organic-based photovoltaics: toward low-cost power generation, MRS Bull. 30 (2005) 10–19. https://doi.org/10.1557/mrs2005.2

[13] E. Kymakis, G.A.J. Amaratunga, Single-wall carbon nanotube/conjugated polymer photovoltaic devices, Appl. Phys. Lett. 80 (2002) 112–114. https://doi.org/10.1063/1.1428416

[14] J.C. Charlier, M. Terrones, M. Baxendale, V. Meunier, T. Zacharia, N.L. Rupesinghe, W.K. Hsu, N. Grobert, H. Terrones, G.A.J. Amaratunga, Enhanced

electron field emission in b-doped carbon nanotubes, Nano Lett. 2 (2002) 1191–1195. https://doi.org/10.1021/nl0256457

[15]   P. Dharap, Z. Li, S. Nagarajaiah, E. V. Barrera, Nanotube film based on single-wall carbon nanotubes for strain sensing, Nanotechnology. 15 (2004) 379–382. https://doi.org/10.1088/0957-4484/15/3/026

[16]   J.U. Lee, Photovoltaic effect in ideal carbon nanotube diodes, Appl. Phys. Lett. 87 (2005) 1–4. https://doi.org/10.1063/1.2010598

[17]   B.O.& M. Gratzelt, Institute, A low-cost, high-efficiency solar cell based on dye-sensitized colloidal TiO_2 films, Nature. 353 (1991) 737–740.

[18]   E. Lancelle-Beltran, P. Prené, C. Boscher, P. Belleville, P. Buvat, C. Sanchez, All-solid-state dye-sensitized nanoporous TiO_2 hybrid solar cells with high energy-conversion efficiency, Adv. Mater. 18 (2006) 2579–2582. https://doi.org/10.1002/adma.200502023

[19]   J.E. Moser, Solar cells: Later rather than sooner, Nat. Mater. 4 (2005) 723–724. https://doi.org/10.1038/nmat1504

[20]   G. Dennler, M.C. Scharber, T. Ameri, P. Denk, K. Forberich, C. Waldauf, C.J. Brabec, Design rules for donors in bulk-heterojunction tandem solar cells-towards 15 % energy-conversion efficiency, Adv. Mater. 20 (2008) 579–583. https://doi.org/10.1002/adma.200702337

[21]   T. Kim, Y.S. Kim, J.Y. Choi, J.H. Jeon, W.W. Park, S.W. Moon, S.M. Kim, S. Han, B. Kim, D.K. Lee, H. Kim, J.Y. Kim, M.J. Ko, K. Kim, Reversed organic-inorganic hybrid tandem solar cells for improved interfacial series resistances and balanced photocurrents, Synthetic Met. 175 (2013) 103–107. https://doi.org/10.1016/j.synthmet.2013.05.002

[22]   Y. Jung, X. Li, N.K. Rajan, A.D. Taylor, M.A. Reed, Record high efficiency single-walled carbon nanotube/silicon p - N junction solar cells, Nano Lett. 13 (2013) 95–99. https://doi.org/10.1021/nl3035652

[23]   M.W. Khan, X. Zuo, Q. Yang, H. Tang, K. Mehmood, U. Rehman, M. Wu, G. Li, multiwall carbon nanotubes boost the short-circuit current of Ru(II) based dye-sensitized solar cells, Nanoscale 12 (2020) 1046–1060. https://doi.org/10.1039/c9nr09227g

[24]   E. Singh, K.S. Kim, G.Y. Yeom, H.S. Nalwa, Two-dimensional transition metal dichalcogenide-based counter electrodes for dye-sensitized solar cells, RSC Adv. 7 (2017) 28234–28290. https://doi.org/10.1039/c7ra03599c

[25]   B. Pradhan, S.K. Batabyal, A.J. Pal, Functionalized carbon nanotubes in donor/acceptor-type photovoltaic devices, Appl. Phys. Lett. 88 (2006) 2–5. https://doi.org/10.1063/1.2179372

[26]   P.R. Somani, S.P. Somani, M. Umeno, Application of metal nanoparticles decorated carbon nanotubes in photovoltaics, Appl. Phys. Lett. 93 (2008) 2006–2009. https://doi.org/10.1063/1.2963470

[27]   P.R. Somani, S.P. Somani, M. Umeno, Application of metal nanoparticles decorated carbon nanotubes in photovoltaics, Appl. Phys. Lett. 93 (2008). https://doi.org/10.1063/1.2963470

[28]   S. Zhang, C. Ji, Z. Bian, P. Yu, L. Zhang, D. Liu, E. Shi, Y. Shang, H. Peng, Q. Cheng, D. Wang, C. Huang, A. Cao, Porous, platinum nanoparticle-adsorbed carbon nanotube yarns for efficient fiber solar cells, ACS Nano. 6 (2012) 7191–7198. https://doi.org/10.1021/nn3022553

[29]   A.A. Arbab, K.C. Sun, I.A. Sahito, M.B. Qadir, Y.S. Choi, S.H. Jeong, A novel activated-charcoal-doped multiwalled carbon nanotube hybrid for quasi-solid-state dye-sensitized solar cell outperforming Pt electrode, ACS Appl. Mater. Interfaces. 8 (2016) 7471–7482. https://doi.org/10.1021/acsami.5b09319

[30]   P.R. Pudasaini, F. Ruiz-zepeda, M. Sharma, D. Elam, A. Ponce, A.A. Ayon, High efficiency hybrid silicon nanopillar-polymer solar cells, Appl. Mater. Interfaces. 5 (2013) 9620–9627. https://doi.org/10.1021/am402598j

[31]   S. Dayal, N. Kopidakis, D.C. Olson, D.S. Ginley, G. Rumbles, Photovoltaic devices with a low band gap polymer and CdSe nanostructures exceeding 3% efficiency, Nano Lett. 10 (2010) 239–242. https://doi.org/10.1021/nl903406s

[32]   V. Kaltenhauser, T. Rath, M. Edler, A. Reichmann, G. Trimmel, Exploring polymer/nanoparticle hybrid solar cells in tandem architecture, RSC Adv. 3 (2013) 18643–18650. https://doi.org/10.1039/c3ra43842b

[33]   J.Y. Kim, P. Vincent, J. Jang, M.S. Jang, M. Choi, J.H. Bae, C. Lee, H. Kim, Versatile use of ZnO interlayer in hybrid solar cells for self-powered near infra-red photo-detecting application, J. Alloys Compd. 813 (2020) 1–7. https://doi.org/10.1016/j.jallcom.2019.152202

[34]   R. Liu, J. Wang, T. Sun, M. Wang, C. Wu, H. Zou, T. Song, X. Zhang, S.T. Lee, Z.L. Wang, B. Sun, Silicon nanowire/polymer hybrid solar cell-supercapacitor: a self-charging power unit with a total efficiency of 10.5%, Nano Lett. 17 (2017) 4240–4247. https://doi.org/10.1021/acs.nanolett.7b01154

[35] B.R. Saunders, M.L. Turner, Nanoparticle-polymer photovoltaic cells, Adv. Colloid Interface Sci. 138 (2008) 1–23. https://doi.org/10.1016/j.cis.2007.09.001

[36] M.H. Yun, J.W. Kim, S.Y. Park, D.S. Kim, B. Walker, J.Y. Kim, High-efficiency, hybrid Si/C60 heterojunction solar cells, J. Mater. Chem. A. 4 (2016) 16410–16417. https://doi.org/10.1039/c6ta02248k

[37] S. Yodyingyong, X. Zhou, Q. Zhang, D. Triampo, J. Xi, K. Park, B. Limketkai, G. Cao, Enhanced photovoltaic performance of nanostructured hybrid solar cell using highly oriented TiO_2 nanotubes, J. Phys. Chem. C. 114 (2010) 21851–21855. https://doi.org/10.1021/jp1077888

[38] N. Balis, V. Dracopoulos, E. Stathatos, N. Boukos, P. Lianos, A solid-state hybrid solar cell made of nc-TiO_2, CdS quantum dots, and P3HT with 2-amino-1-methylbenzimidazole as an interface modifier, J. Phys. Chem. C. 115 (2011) 10911–10916. https://doi.org/10.1021/jp2022264

[39] B. Reeja-Jayan, A. Manthiram, Understanding the improved stability of hybrid polymer solar cells fabricated with copper electrodes, ACS Appl. Mater. Interfaces. 3 (2011) 1492–1501. https://doi.org/10.1021/am200067d

[40] W.P. Liao, S.C. Hsu, W.H. Lin, J.J. Wu, Hierarchical TiO_2 nanostructured array/P3HT hybrid solar cells with interfacial modification, J. Phys. Chem. C. 116 (2012) 15938–15945. https://doi.org/10.1021/jp304915x

[41] S. Abdulalmohsin, J.B. Cui, Graphene-enriched P3HT and porphyrin-modified ZnO nanowire arrays for hybrid solar cell applications, J. Phys. Chem. C. 116 (2012) 9433–9438. https://doi.org/10.1021/jp301881s

[42] M.J. Greaney, S. Das, D.H. Webber, S.E. Bradforth, R.L. Brutchey, Improving open circuit potential in hybrid P3HT: CdSe bulk heterojunction solar cells via colloidal tert-butylthiol ligand exchange, ACS Nano. 6 (2012) 4222–4230. https://doi.org/10.1021/nn3007509

[43] S. Jeong, E.C. Garnett, S. Wang, Z. Yu, S. Fan, M.L. Brongersma, M.D. McGehee, Y. Cui, Hybrid silicon nanocone-polymer solar cells, Nano Lett. 12 (2012) 2971–2976. https://doi.org/10.1021/nl300713x

[44] S.H. Eom, M.J. Baek, H. Park, L. Yan, S. Liu, W. You, S.H. Lee, Roles of interfacial modifiers in hybrid solar cells: Inorganic/polymer bilayer vs inorganic/polymer:Fullerene bulk heterojunction, ACS Appl. Mater. Interfaces. 6 (2014) 803–810. https://doi.org/10.1021/am402684w

[45]   Y. Qin, Y. Cheng, L. Jiang, X. Jin, M. Li, X. Luo, G. Liao, T. Wei, Q. Li, Top-down strategy toward versatile graphene quantum dots for organic/inorganic hybrid solar cells, ACS Sustain. Chem. Eng. 3 (2015) 637–644. https://doi.org/10.1021/sc500761n

[46]   Y.T. Lee, F.R. Lin, C.H. Chen, Z. Pei, A 14.7% organic/silicon nanoholes hybrid solar cell via interfacial engineering by solution-processed inorganic conformal layer, ACS Appl. Mater. Interfaces. 8 (2016) 34537–34545. https://doi.org/10.1021/acsami.6b10741

[47]   H.D. Um, D. Choi, A. Choi, J.H. Seo, K. Seo, Embedded metal electrode for organic-inorganic hybrid nanowire solar cells, ACS Nano. 11 (2017) 6218–6224. https://doi.org/10.1021/acsnano.7b02322

[48]   D.H. Shin, J.H. Kim, S.H. Choi, High-performance conducting polymer/Si nanowires hybrid solar cells using multilayer-graphene transparent conductive electrode and back surface passivation layer, ACS Sustain. Chem. Eng. 6 (2018) 12446–12452. https://doi.org/10.1021/acssuschemeng.8b03005

[49]   M.D. Brown, T. Suteewong, R.S.S. Kumar, V.D. Innocenzo, A. Petrozza, M.M. Lee, U. Wiesner, H.J. Snaith, Plasmonic dye-sensitized solar cells using core - shell metal - insulator nanoparticles, Nano Lett. 11 (2011) 438–445. https://doi.org/10.1021/nl1031106

[50]   F. Wang, N.K. Subbaiyan, Q. Wang, C. Rochford, G. Xu, R. Lu, A. Elliot, F. D'souza, R. Hui, J. Wu, Development of nanopatterned fluorine-doped tin oxide electrodes for dye-sensitized solar cells with improved light trapping, ACS Appl. Mater. Interfaces. 4 (2012) 1565–1572. https://doi.org/10.1021/am201760q

[51]   M.K. Gangishetty, K.E. Lee, R.W.J. Scott, T.L. Kelly, Plasmonic enhancement of dye sensitized solar cells in the red-to-near-infrared region using triangular core-shell $Ag@SiO_2$ nanoparticles, ACS Appl. Mater. Interfaces. 5 (2013) 11044–11051. https://doi.org/10.1021/am403280r

[52]   S. Kelkar, K. Pandey, S. Agarkar, N. Saikhedkar, M. Tathavadekar, I. Agrawal, R.V.N. Gundloori, S. Ogale, Functionally engineered egg albumen gel for quasi-solid dye sensitized solar cells, ACS Sustain. Chem. Eng. 2 (2014) 2707–2714. https://doi.org/10.1021/sc5004488

[53]   C. Karam, R. Habchi, S. Tingry, P. Miele, M. Bechelany, Design of multilayers of urchin-like ZnO nanowires coated with TiO_2 nanostructures for dye-sensitized solar cells , ACS Appl. Nano Mater. 1 (2018) 3705–3714. https://doi.org/10.1021/acsanm.8b00849

Materials Research Forum LLC
https://doi.org/10.21741/9781644901410-7

[54]   J.S. Kang, J. Kim, J.S. Kim, K. Nam, H. Jo, Y.J. Son, J. Kang, J. Jeong, H. Choe, T.H. Kwon, Y.E. Sung, Electrochemically synthesized mesoscopic nickel oxide films as photocathodes for dye-sensitized solar cells, ACS Appl. Energy Mater. 1 (2018) 4178–4185. https://doi.org/10.1021/acsaem.8b00834

[55]   F.A. AL-Temimei, A.H. OmranAlkhayatt, A DFT/TD-DFT investigation on the efficiency of new dyes based on ethyl red dye as a dye-sensitized solar cell light-absorbing material, Optik (Stuttg). (2019) 163920. https://doi.org/10.1016/j.ijleo.2019.163920

[56]   Y. Liu, L.A. Renna, M. Bag, Z.A. Page, P. Kim, J. Choi, T. Emrick, D. Venkataraman, T.P. Russell, High efficiency tandem thin-perovskite/polymer solar cells with a graded recombination layer, ACS Appl. Mater. Interfaces. 8 (2016) 7070–7076. https://doi.org/10.1021/acsami.5b12740

[57]   L. Kranz, A. Abate, T. Feurer, F. Fu, E. Avancini, J. Löckinger, P. Reinhard, S.M. Zakeeruddin, M. Grätzel, S. Buecheler, A.N. Tiwari, High-efficiency polycrystalline thin film tandem solar cells, J. Phys. Chem. Lett. 6 (2015) 2676–2681. https://doi.org/10.1021/acs.jpclett.5b01108

[58]   A. Guchhait, H.A. Dewi, S.W. Leow, H. Wang, G. Han, F. Bin Suhaimi, S. Mhaisalkar, L.H. Wong, N. Mathews, Over 20% efficient CIGS-perovskite tandem solar cells, ACS Energy Lett. 2 (2017) 807–812. https://doi.org/10.1021/acsenergylett.7b00187

[59]   G. Li, H. Li, J.Y.L. Ho, M. Wong, H.S. Kwok, Nanopyramid structure for ultrathin c-Si tandem solar cells, Nano Lett. 14 (2014) 2563–2568. https://doi.org/10.1021/nl500366c

[60]   K.T. Vansant, J. Simon, J.F. Geisz, E.L. Warren, K.L. Schulte, A.J. Ptak, M.S. Young, M. Rienäcker, H. Schulte-Huxel, R. Peibst, A.C. Tamboli, Toward Low-cost 4-terminal GaAs//Si tandem solar cells, ACS Appl. Energy Mater. 2 (2019) 2375–2380. https://doi.org/10.1021/acsaem.9b00018

Keyword Index

About the Editors

Dr. Inamuddin is working as Assistant Professor at the Department of Applied Chemistry, Aligarh Muslim University, Aligarh, India. He obtained Master of Science degree in Organic Chemistry from Chaudhary Charan Singh (CCS) University, Meerut, India, in 2002. He received his Master of Philosophy and Doctor of Philosophy degrees in Applied Chemistry from Aligarh Muslim University (AMU), India, in 2004 and 2007, respectively. He has extensive research experience in multidisciplinary fields of Analytical Chemistry, Materials Chemistry, and Electrochemistry and, more specifically, Renewable Energy and Environment. He has worked on different research projects as project fellow and senior research fellow funded by University Grants Commission (UGC), Government of India, and Council of Scientific and Industrial Research (CSIR), Government of India. He has received Fast Track Young Scientist Award from the Department of Science and Technology, India, to work in the area of bending actuators and artificial muscles. He has completed four major research projects sanctioned by University Grant Commission, Department of Science and Technology, Council of Scientific and Industrial Research, and Council of Science and Technology, India. He has published 176 research articles in international journals of repute and nineteen book chapters in knowledge-based book editions published by renowned international publishers. He has published 115 edited books with Springer (U.K.), Elsevier, Nova Science Publishers, Inc. (U.S.A.), CRC Press Taylor & Francis Asia Pacific, Trans Tech Publications Ltd. (Switzerland), IntechOpen Limited (U.K.), Wiley-Scrivener, (U.S.A.) and Materials Research Forum LLC (U.S.A). He is a member of various journals' editorial boards. He is also serving as Associate Editor for journals (Environmental Chemistry Letter, Applied Water Science and Euro-Mediterranean Journal for Environmental Integration, Springer-Nature), Frontiers Section Editor (Current Analytical Chemistry, Bentham Science Publishers), Editorial Board Member (Scientific Reports-Nature), Editor (Eurasian Journal of Analytical Chemistry), and Review Editor (Frontiers in Chemistry, Frontiers, U.K.) He is also guest-editing various special thematic special issues to the journals of Elsevier, Bentham Science Publishers, and John Wiley & Sons, Inc. He has attended as well as chaired sessions in various international and national conferences. He has worked as a Postdoctoral Fellow, leading a research team at the Creative Research Initiative Center for Bio-Artificial Muscle, Hanyang University, South Korea, in the field of renewable energy, especially biofuel cells. He has also worked as a Postdoctoral Fellow at the Center of Research Excellence in Renewable Energy, King Fahd University of Petroleum and Minerals, Saudi Arabia, in the field of polymer electrolyte membrane fuel cells and computational fluid dynamics of polymer electrolyte membrane fuel cells. He is a life member of the Journal of the Indian

Chemical Society. His research interest includes ion exchange materials, a sensor for heavy metal ions, biofuel cells, supercapacitors and bending actuators.

Dr. Tauseef Ahmad Rangreez is working as a postdoctoral fellow at National Institute of Technology, Srinagar, India. He completed his Ph.D in Applied Chemistry, from Aligarh Muslim University, Aligarh, India on the topic "Development of Nanostructure Organic-Inorganic Composite Materials based Sensors for Inorganic Pollutants". He worked as a Project Fellow under the UGC Funded Research Project entitled "Development of Nanostructured Conductive Organic Inorganic Composite Materials based sensors Functionalities for Organic and Inorganic Pollutants". He completed his Masters in Chemistry from Jamia Hamdard, New Delhi. He has published several research articles of international repute. He has edited books with Springer and Materials Science Forum LLC, U.S.A. His research interest includes ion exchange chromatography, development of nanocomposite sensors for heavy metals and biosensors.

Dr. Mohd Imran Ahamed received his Ph.D degree on the topic "Synthesis and characterization of inorganic-organic composite heavy metals selective cation-exchangers and their analytical applications", from Aligarh Muslim University, Aligarh, India in 2019. He has published several research and review articles in the journals of international recognition. Springer (U.K.), Elsevier, CRC Press Taylor & Francis Asia Pacific and Materials Research Forum LLC (U.S.A). He has completed his B.Sc. (Hons) Chemistry from Aligarh Muslim University, Aligarh, India, and M.Sc. (Organic Chemistry) from Dr. Bhimrao Ambedkar University, Agra, India. He has co-edited more than 20 books with Springer (U.K.), Elsevier, CRC Press Taylor & Francis Asia Pacific and Materials Research Forum LLC (U.S.A) and Wiley-Scrivener, (U.S.A.). His research work includes ion-exchange chromatography, wastewater treatment, and analysis, bending actuator and electrospinning.

Dr. Hamida-Tun-Nisa Chisti is currently working as an Associate Professor in the Department of Chemistry, National Institute of Technology Srinagar, India. She has received her B.Sc. Hons., Masters and Ph.D. in Chemistry from Aligarh Muslim University, Aligarh, and later has joined her services as Lecturer in 2008 at NIT Srinagar. Her research focus is in the fields of Inorganic Chemistry, Environmental Chemistry, and Material Chemistry. She has published several research articles in international journals of repute. She has authored 4 books and 6 book chapters. She is a Member Royal Society of Chemistry (MRSC), ACS and also a life member of Asian Polymer Association, Indian Council of Chemists, Life member of the International Association of Engineers (**IAENG**), Hong Kong, Life member of Eco Ethics

International Union (**EEIU**), Germany and many more. She is also serving as a reviewer for many journals.

www.ingramcontent.com/pod-product-compliance
Lightning Source LLC
Chambersburg PA
CBHW071228210326
41597CB00016B/1987

* 9 7 8 1 6 4 4 9 0 1 4 0 3 *